Science and the politics of openness

Manchester University Press

Science and the politics of openness

Here be monsters

Edited by Brigitte Nerlich, Sarah Hartley, Sujatha Raman and Alexander Smith

Manchester University Press

Copyright © Manchester University Press 2018

While copyright in the volume as a whole is vested in Manchester University Press, copyright in individual chapters belongs to their respective authors, and no chapter may be reproduced wholly or in part without the express permission in writing of both author and publisher.

Published by Manchester University Press
Altrincham Street, Manchester M1 7JA

www.manchesteruniversitypress.co.uk

British Library Cataloguing-in-Publication Data
A catalogue record for this book is available from the British Library

ISBN 978 1 5261 0646 9 hardback
ISBN 978 1 5261 0647 6 open access

First published 2018

An electronic version of this book is also available under a Creative Commons (CC-BY-NC-ND) licence.

The publisher has no responsibility for the persistence or accuracy of URLs for any external or third-party internet websites referred to in this book, and does not guarantee that any content on such websites is, or will remain, accurate or appropriate.

Typeset
by Toppan Best-set Premedia Limited
Printed in Great Britain
by CPI Group (UK) Ltd, Croydon, CR0 4YY

Contents

List of figures and tables	*page* viii
List of contributors	ix

Introduction – Brigitte Nerlich, Sujatha Raman, Sarah Hartley, Alexander Thomas T. Smith — 1

Coda: reflections on the politics of openness in a new world order – Alexander Thomas T. Smith — 12

Part I Transparency

1. Transparency – Benjamin Worthy — 23
2. Open access: the beast that no-one could – or should – control? – Stephen Curry — 33
3. Assuaging fears of monstrousness: UK and Swiss initiatives to open up animal laboratory research – Carmen M. McLeod — 55
4. What counts as evidence in adjudicating asylum claims? Locating the monsters in the machine: an investigation of faith-based claims – Roda Madziva and Vivien Lowndes — 75

Part II Responsibility

5. Responsibility – Barbara Prainsack and Sabina Leonelli — 97

6   Leviathan and the hybrid network: Future Earth,
    co-production and the experimental life of a global
    institution – Eleanor Hadley Kershaw   107

7   'Opening up' energy transitions research for
    development – Alison Mohr   131

8   Monstrous regiment versus Monsters Inc.:
    competing imaginaries of science and social
    order in responsible (research and) innovation
    – Stevienna de Saille and Paul Martin   148

Part III Expertise

9   Expertise – Mark B. Brown   169

10 Disentangling risk assessment: new roles for
    experts and publics – Sarah Hartley and
    Adam Kokotovich   176

11 Monstrous materialities: ash dieback and
    plant biosecurity in Britain – Judith Tsouvalis   195

12 *An Inconvenient Truth*: a social representation
    of scientific expertise – Warren Pearce and
    Brigitte Nerlich   212

13 'Science Matters' and the public interest:
    the role of minority engagement – Sujatha Raman,
    Pru Hobson-West, Mimi E. Lam and Kate Millar   230

Part IV Faith

14 Faith – Chris Toumey   253

15 Re-examining 'creationist' monsters in the uncharted
    waters of social studies of science and religion – Fern
    Elsdon-Baker   259

16 Playing God: religious influences on the depictions
    of science in mainstream movies – David A. Kirby
    and Amy C. Chambers   278

Afterword: monstrous markets – neo-liberalism,
populism and the demise of the public university
– John Holmwood and Jan Balon   303

Epilogue: publics, hybrids, transparency, monsters and the
changing landscape around science – Stephen Turner   322

Index   333

Figures and tables

Figure

16.1 Pamphlet on the creationist origins of
Jurassic Park's dinosaurs disseminated by
the Southwest Radio Church. 292

Table

10.1 A framework for public involvement in
risk assessment 189

Contributors

Jan Balon is Lecturer at the Institute of Sociological Studies, Charles University in Prague and Head of Department and Researcher at the Centre for Science, Technology and Society Studies, Czech Academy of Sciences. With Marek Skovajsa, he is author of *Sociology in the Czech Republic: Between East and West* (Palgrave, 2017).

Mark B. Brown is professor in the Department of Political Science at California State University, Sacramento. He is the author of *Science in Democracy: Expertise, Institutions, and Representation* (MIT Press, 2009), and various publications on the politics of expertise, political representation, bioethics, climate change and related topics.

Amy C. Chambers is a research associate in science communication and screen studies at Newcastle University, working on an AHRC-funded project that seeks to map the history of imagined futures in speculative fiction. Her publications examine the intersection of science, religion, and entertainment media, and also science fiction and horror film/TV.

Stephen Curry is a professor of structural biology at Imperial College London. An active blogger and campaigner, he writes regularly in the *Guardian* and elsewhere about open access and the larger social responsibilities of scientists. Stephen is a member of the board of both Science is Vital and the Campaign for Science and Engineering.

Stevienna de Saille is currently a research fellow in the Institute for the Study of the Human at the University of Sheffield, studying genome editing, the human-tissue bioeconomy and responsible innovation through the lens of heterodox economics. She was a research fellow on the Leverhulme Trust project Making Science Public, studying the emergence of responsible research and innovation.

Fern Elsdon-Baker is a professor and the Director of the Centre for Science, Knowledge and Belief in Society at Newman University, Birmingham, and the principal investigator on the Science and Religion: Exploring the Spectrum project. Previously she was head of the British Council's Darwin Now project, which ran in fifty countries worldwide.

Eleanor Hadley Kershaw is a doctoral researcher in science and technology studies at the Institute for Science and Society, University of Nottingham. Her PhD is funded by the Leverhulme Trust as part of the Making Science Public programme. Her background is in English literature, theatre/performance, sustainability, environment and international science policy.

Sarah Hartley is a senior lecturer at the University of Exeter Business School, where she researches science, technology and innovation governance. Her current work examines the role of publics and experts – in the development of genetically modified insects, in gene drive and in genome editing – in areas of global health, sustainable agriculture and food security.

Pru Hobson-West has expertise in medical sociology, science and technology studies, and ethics. Her work focuses on controversies such as childhood vaccination and animal experimentation. Pru also leads work in the emerging field of veterinary sociology. She is based at the Centre for Applied Bioethics in the School of Veterinary Medicine and Science at the University of Nottingham.

John Holmwood is Professor of Sociology at the University of Nottingham. He is the founder of the Campaign for the Public University. He blogs regularly on higher-education issues for the Campaign for the Public University, Research Blogs, Open Democracy and other venues. His current research addresses issues of pragmatism and public sociology. He co-edits the free online magazine *Discover Society*.

David A. Kirby is Senior Lecturer in Science Communication Studies at the University of Manchester. His book *Lab Coats in Hollywood*

(MIT Press, 2013) examines the collaborations between scientists and the entertainment industry. He is currently writing the book *Indecent Science*, which will explore the historic interactions between science, religion and movie censorship.

Adam Kokotovich is a postdoctoral research associate at the University of Minnesota, studying invasive-species risk assessment and management. An interdisciplinary social scientist, he is interested in highlighting and opening to reflexive scrutiny the assumptions that inform decision making related to science, risk and the environment.

Mimi E. Lam is a Marie Curie Fellow at the University of Bergen, Centre for the Study of the Sciences and the Humanities. She specialises in seafood ethics and sustainability, marine governance, and the science–policy interface. She studies values, beliefs and perceptions of nature to inform societal decision making for a sustainable and just future.

Sabina Leonelli is Professor in Philosophy and History of Science and Co-Director of the Exeter Centre for the Study of the Life Sciences at the University of Exeter. Her current research focuses on the philosophy, history and sociology of data-intensive science, especially the research processes, scientific outputs and social embedding of open science, open data and big data.

Vivien Lowndes is Professor of Public Policy at the University of Birmingham. She was Deputy Chair of the Politics and International Studies Sub-Panel for the 2014 Research Excellence Framework. Vivien's research interests focus on local governance and community engagement. She is author (with Mark Roberts) of *Why Institutions Matter* (Palgrave, 2013).

Roda Madziva is an Assistant Professor of Sociology at the University of Nottingham. Her research focuses on understanding the association between immigration policies and migrants' lived experiences. She has worked on the use of evidence in immigration policy as part of a Leverhulme-funded programme, Making Science Public: Opportunities and Challenges. She has worked closely with migrant support organisations, faith-based communities and churches at both local and national level.

Paul Martin is Professor of Sociology of Science and Technology in the Department of Sociological Studies, University of Sheffield. He recently completed a project for the Making Science Public programme (with Stevienna de Saille) on ideas of responsible innovation, and is currently working on epigenetics and the role of science in policymaking.

Carmen McLeod is an anthropologist and postdoctoral researcher based at the University of Oxford in the School of Geography and the Environment, where she is exploring the social dimensions of the human microbiome. Carmen was a research fellow (February 2013 to June 2015) on the Leverhulme Trust Programme Making Science Public, working on the project Animals and the Making of Scientific Knowledge.

Kate Millar is Director of the Centre for Applied Bioethics, University of Nottingham, and President of the European Society for Agricultural and Food Ethics (EurSafe). Kate's research focuses on bioethics, with a particular interest in agri-food and veterinary ethics, development of ethical frameworks (e.g. Ethical Matrix), and stakeholder participation methods.

Alison Mohr is Co-Director (Research) at the Institute for Science and Society in the School of Sociology and Social Policy, University of Nottingham. Alison's research explores the social, cultural, political and environmental dimensions of energy transitions. She sits on the management boards of the university-wide Energy Technologies Research Institute and Energy Research Priority Area to champion interdisciplinary research and public and policy engagement.

Brigitte Nerlich is Emeritus Professor of Science, Language and Society at the University of Nottingham. She has published widely on semantics, pragmatics, cognitive linguistics, and, more recently, the sociology of health and illness and the cultural study of science and technology. She was Director of the Leverhulme Making Science Public programme between 2012 and 2016.

Warren Pearce is Faculty Fellow (iHuman) within the Department of Sociological Studies, University of Sheffield. He researches the relationship between science, policy and publics, with a current focus on two main themes: public inclusion in research governance, and climate-change communication and policy.

Barbara Prainsack is a professor in the Department of Political Science at the University of Vienna, Austria, and in the Department of Global Health and Social Medicine at King's College London. Her work addresses the social, regulatory and ethical dimensions of biomedicine and the biosciences. Her current projects focus on personalised and 'precision' medicine and the role of solidarity in guiding practice and policy in biomedicine.

Sujatha Raman is Associate Professor in Science and Technology Studies and Co-Director (Research) at the Institute for Science and Society (ISS) in the School of Sociology and Social Policy, University of Nottingham. She was Deputy Director of the Leverhulme Making Science Public research programme (2012–2016) and, latterly, the Director (2016–2018). Her research focuses on science, democracy and sustainability.

Alexander Thomas T. Smith holds a PhD in social anthropology from Edinburgh University and teaches sociology at Warwick. He conducts research on conservatism, local politics and religion in Scotland and the USA. He is the author of *Devolution and the Scottish Conservatives* (Manchester University Press, 2011) and the editor (with John Holmwood) of *Sociologies of Moderation* (Wiley Blackwell, 2013). Alex's next book, *Democracy Begins at Home: Political Moderation in Red State America*, will be published by the University Press of Kansas.

Chris Toumey is a cultural anthropologist who works in the anthropology of science. Since 2013 he has focused on societal and cultural issues in nanotechnology. He is author or co-author of more than eighty papers on nanotech, including his commentaries that appear four times a year in *Nature Nanotechnology*. Based on that work, plus earlier work on issues of science and religion, he is especially interested in relations between technology and religion.

Judith Tsouvalis is a geographer with research interests in the making and politics of knowledge on nature and the non-human; risk perception; power relations at the interface of science, policymaking, politics and publics; new materialism; and political ecology. Her empirical work has focused on forestry, farming, water management, plant biosecurity, science advice and public participation.

Stephen Turner is Distinguished University Professor of Philosophy at the University of South Florida. His writings include *Liberal Democracy 3.0: Civil Society in an Age of Experts*, and *The Politics of Expertise* (both Sage, 2003). He has written extensively on the history of geology, statistics and social science.

Benjamin Worthy is a lecturer in politics at Birkbeck College, University of London. His specialisms include political leadership and government transparency. He has written extensively on freedom of information and transparency, and his new monograph examines *The Politics of Freedom of Information: How and Why Governments Pass Laws that Threaten Their Power* (Manchester University Press). He is also researching the impact of the UK Government's transparency agenda.

Introduction

Brigitte Nerlich, Sujatha Raman, Sarah Hartley, Alexander Thomas T. Smith

In recent years the relation between science and society has become strained. In some parts of the world, mainly in the United States, science is said to be 'at war' with society (Otto, 2016). In others, particularly the United Kingdom, scientists have been dragged into debates over suspicion and contempt of experts, primarily economists (Mance, 2016). These developments play out against a series of crises in science, technology, politics and the economy, which are all interlinked.

In politics and economics, one can mention the 2008 financial crisis, threats posed by terrorism, rising tensions around the place of religion in science and society, the ascent of political populism, and, more recently, debates about the United Kingdom leaving the European Union – an exit (dubbed 'Brexit') that will have profound consequences for society, science and different forms of expertise. In the United States, a new presidency challenges established relations between science and politics, as well as between these domains, journalism and the public. In technology, controversies arise from increasing digitalisation, automation, cybercrime and much more, but also from the maturing of novel energy technologies and biotechnologies. These new technologies bring with them a range of ethical problems.

This volume uses recent developments in science–society relations as a focal point for exploring the tensions and contradictions raised by these large-scale issues. Over the last thirty years in the UK and beyond, science, policy and the public have come into conflict in a series of political crises. Examples include the BSE (bovine spongiform encephalopathy) or 'mad cow disease' crisis, the almost intractable disputes about the commercialisation of genetically modified (GM) foods and crops, the effects of scientific dissent on public health (e.g. the measles, mumps and rubella (MMR) vaccine) and the impacts of the politicisation of climate science on climate policy ('Climategate'). Since then, we have seen the emergence of new controversies involving the use of scientific evidence in policy (e.g. on the proposed role of dental evidence to assess the age of child refugees; the emergence of ash dieback disease) and on the conduct of research and innovation (e.g. on gene editing; management of the Zika virus), where issues related to security are ever-present. Last but not least, many of the topics listed above are global in nature. In this context we have seen the emergence of highly public conflicts around global institutions, such as the Intergovernmental Panel on Climate Change, where science and politics meet.

In order to deal with such emerging and enduring matters and their impacts on trust in science and politics, policymakers have proposed various solutions, such as promoting greater public engagement with science and policy, co-design of scientific research with stakeholders, open and participatory forms of innovation, increasing transparency in scientific advice for policymaking, and enhancing open access to scientific data and research outputs. Such solutions have begun to exploit a number of new digital technologies and algorithms which can be used for good or for ill – for sharing information quickly; for making information public and available for public scrutiny; or for increasing surveillance by the state, social groups or oneself.

We live in a society that increasingly aspires to open access, open data, open science and open policy. Underpinning this trend is an assumption that transparency and openness are in the 'public interest'. They are, but there may well be dark sides to this trend that need to be examined. Indeed, these solutions to scientific and political crises might conceal or provoke a number of problems, challenges and questions – some new, some timeless. These are the 'monsters' hiding behind transparency, publicness and openness that we want to track down in this book. Keeping in mind that unknown troubles might lurk behind apparent solutions, we ask: what does making science more public, open or transparent mean, in theory and practice? Who are 'the public' and how are they constituted? What might 'public science' mean for the authority and independence of science and the capacity of publics to engage with science? What are the political implications of making science more public or transparent, and how does this relate to issues of legitimacy and transparency in politics and policymaking? What role do interested citizens play in the creation of science and the making of science policy? Who controls the new technologies and enterprises of openness and transparency? And what will happen in the future, given radical changes that are happening in science and technology, as well as politics and policy, globally and nationally?

Trying to find answers to these questions provides us with a much-needed opportunity to rethink the relationship between science and politics and, more importantly, the role of science in public, the role of publics in science, and the role of expertise in science and policymaking, as well the role of faith in science and society. Others have examined these issues, but we seek to put them in conversation with wider political developments around migration, religion and neo-liberalism.

The book

The chapters in this volume are based on work carried out within the Leverhulme-Trust-funded Making Science Public research programme (2012–2017), which explored the relationships between science, politics and publics through a number of topical case studies. The chapters challenge received wisdoms about openness and transparency and highlight and map the pitfalls and dangers – the 'monsters' in openness and transparency. The book is motivated by the sense that there might be metaphorical dragons or monsters hiding behind policy initiatives to 'open up' science in response to perceived legitimacy crises in research and innovation systems and in the relationship between science and policymaking.

The phrase 'here be monsters' or 'here be dragons' is commonly believed to have been used on ancient maps to indicate unexplored territories which might hide unknown beasts. Etymologically, the figure of the monster is double-edged and ambiguous in a way that invites reflection (Haraway, 1992). 'Monster' has twin meanings: the monster serves to both *warn* (Latin: *monere*) and to *show* (*monstrare*). Calls and efforts to open up science evoke multiple conflicting imperatives, hopes and anxieties, which we explore.

Where earlier works in science and technology studies employed the monster or golem figure to study the social aspects of science and its technological creations (Collins and Pinch, 1993; Haraway, 1992; Law, 1991), we argue that the time is ripe to examine the positive and negative effects of contemporary policy initiatives and institutions which purport to bring science, society and publics closer together through processes of openness, access and transparency. Developed as solutions to perceived crises in science/society relations, a variety of policy initiatives hide dilemmas that need to be made visible and need to be discussed out in the open. In sum, the chapters in this book explore the unfolding contradictions around efforts to 'make science public'.

Science and the politics of openness: challenges and dilemmas

For many years now, some science and technology studies scholars have called for new forms of 'post-normal' science (Funtowicz and Ravetz, 1993), where scientific claims and their underlying assumptions are opened up to wider scrutiny, allowing a new 'democratisation of science' (Brown, 2009). Sheila Jasanoff (2002) has argued that we are witnessing a so-called 'constitutional moment' in which the claims of scientific knowledge and technology on behalf of the public good need to be openly justified rather than taken for granted. However, promises of openness, transparency and greater engagement generated by these developments must always be critically assessed. 'People may not possess enough specialized knowledge and material resources' to participate in apparently open forums (Jasanoff, 2003: 237). Likewise, privately sponsored industry research can all too easily be used through transparency laws to destabilise public science – for example, in cases of environmental and health-and-safety regulations introduced in the public interest (Jasanoff, 2006) – or to undermine political action on climate change, as journalist Delingpole (2010), for one, has attempted to do.

With the institutionalisation of scientific advice in government and an emphasis on evidence-based policy, scientific evidence has taken centre stage in many public and political conflicts. However, there are long-standing fears of a possible 'scientisation' or 'technocratisation' of politics (Habermas, 1970), which may shut out voices from areas other than science from the policymaking process. In this context, different stakeholders are using a range of experts and cherry-picking the claims or counterclaims that support their own political position. There are enduring debates about the status of expertise, and calls to acknowledge the status of lay experts (Wynne, 1992). These have in turn informed wider debates about opening up and democratising discussions involving scientific knowledge and public policy matters. Science cannot function, argues the political philosopher Stephen Turner (2003), without some monopolisation of expertise. He therefore suggests that it is intrinsically impossible to subject specialised knowledge to democratic discussion (see also Collins, 2014). Yet, wider discussions about the purpose of science and its role in matters of collective interest are political and value based. These cannot, *a priori*, be left to certified experts alone (Sarewitz, 2004), especially when laypeople or rival experts periodically mobilise in particular times and places to call attention to novel dimensions of public importance. In this context, how policymakers or science advisors engage with different forms of expertise becomes crucial.

Scientists have always interacted with a variety of publics, and science communication is as old as science itself. However, scientists are increasingly obliged to get involved in public engagement and, more recently, to practise responsible research and innovation, for reasons other than consulting, communicating with or involving publics. Opportunities for engagement with science and policy are being gradually replaced by mandated activities serving political functions, such as gaining 'impact' and contributing to economic growth. Fostering public engagement with science has become one of many political and economic performance indicators. This may lead to the instrumentalisation of public engagement (Watermeyer, 2012), and a loss of trust in science and scientists.

These developments happen in a context where higher education serves no longer to be primarily a public good but is increasingly marketised and privatised. Universities in the UK are now funded by students (or their parents) who pay tuition fees as consumers rather than as co-producers of knowledge. This means that universities now have a much-weakened relationship with the British public, on whom they no longer depend for the bulk of their teaching budget. There are concerns amongst leading scientists, ethicists and lawyers about 'who owns science' – that is, about the privatisation of science – and calls are being heard for science to serve the public good more explicitly (see Holmwood, 2011; University of Manchester, Institute for Science, Ethics and Innovation, 2010).

Science, politics and publics are entangled in a complex relationship with normative ideas about openness, transparency and publicness, ideals that are being challenged not only by classic science/policy scandals such as those mentioned above (BSE, GM, MMR, Climategate). They are also being challenged by the emergence of new technologies, especially digital technologies envisaged to deliver some of those ideals, and by institutional and technological changes in the spaces where science takes place (universities and industries) and where politics takes place (government, the media, public forums, public consultations, etc.). The fact that political and policy debates, expertise, and the media have been entangled with such normative questions for some time has, in recent decades, become a growing source of instability in Western liberal democracies (see Smith and Holmwood, 2013). This has raised anxieties about the role of expertise and evidence in public debate in an age where countering 'fake news' and political disinformation – or what might have once been described as propaganda or even 'psychological operations' – has become a central concern for policymakers, electoral strategists, journalists, broadcasters and even intelligence agencies following the election of President Donald Trump in the 2016 US elections.

We are now living in a world where words like 'facts' and 'truth' have to be used with considerable care in science and politics, where facts and truths compete with alternative facts or someone else's truth. In an article for the *Financial Times* entitled 'The Problem with Facts', the economist Tim Harford explores, as the subtitle says, 'how today's politicians deal with inconvenient truths'. He uses a phrase that has gained popularity in science and society circles since 2006, when Al Gore produced his notorious film *An Inconvenient Truth*, intended to make what he held to be truths about climate change public. In chapter 12, Warren Pearce and Brigitte Nerlich examine in detail the film and its reverberations through ensuing climate-change debates.

Such debates foreshadowed and even rehearsed arguments that are now raging through most of science and politics, particularly in the USA.

Meanwhile, novel energy technologies encounter issues of justice and fairness when deployed across the world. In chapter 7 Alison Mohr argues that such technologies and innovations might also hide monsters behind their veneer of novelty and service to the agendas of more openness and transparency. The open-access agenda is one major example in the context of research (Holmwood, 2013). The future of new frameworks of responsible research and innovation (RRI) within this wider context is therefore in question. On the one hand, the concept of RRI promises to open up opportunities to rethink the purposes of innovation (Owen et al., 2013) and diversify ways of innovating in response to a societal challenge (Hartley et al., 2016). On the other hand, RRI is becoming reduced to established ways of assessing a specific technology in terms of its risks and benefits. All these developments need to be monitored and scrutinised not only for the opportunities and chances for improvement they offer but also for the pitfalls and contradictions they might contain.

Themes

The chapters in this book map and illuminate issues ('monsters') in specific areas of science/policy practices where the complex problems identified above play out in particular ways and in specific cases.

The book is organised into four parts, around the themes of (1) **transparency** in the context of science in the public sphere; (2) **responsibility** in the context of contemporary research practice and governance, both globally and more locally; (3) **expertise** in the context of policymaking, risk assessment and the regulation of science; and (4) **faith** in the context of emerging tensions and misunderstandings between science, politics and publics regarding issues of religion. Each of the four parts contains an opening essay by an expert on the theme, and the book closes with an afterword and an epilogue reflecting on the contributions to the book.

Transparency

This part opens the book with an exploration of one its core topics; namely, transparency and openness and how they play out within various institutional and policy domains. Three chapters circle these concepts in different ways. **Stephen Curry** deals with an issue that has risen to prominence in science and university research in recent years – 'open access'. He examines not only the potential of open access to break down barriers and open up academic research and knowledge to the wider public, but also the many barriers that exist or are emerging to impede the open-access movement. **Carmen McLeod** deals with issues of transparency and secrecy in the context of animal research through the lens of two transparency initiatives: the Swiss Basel Declaration announced in 2011, and the UK Concordat on Openness in Animal Research launched in 2012. In the final chapter of this part, **Roda Madziva** and **Vivien Lowndes** deal with transparency, evidence and publics in the context of a very topical issue – immigration. This chapter also contributes indirectly to the last part of the book, which deals with faith, as Madziva and Lowndes investigate faith-based claims being used when adjudicating asylum applications.

The part introduction by **Benjamin Worthy** dissects the concepts of transparency and openness and puts them into the context of recent research on these topics, as well as work on related issues around security, privacy, confidentiality and accountability. Worthy also highlights problems with radical openness in a context where 'people' might not be willing, able or interested to make use of the opportunities such openness affords them.

Responsibility

This part continues to explore the topic of openness, but with an additional focus on responsibility and justice. Three chapters move from the global to the more local, and from global environmental change and energy justice to concerns about responsible innovation in the context of Western concerns with genetically modified foods and crops.

Eleanor Hadley Kershaw presents us with an overview of the opportunities and challenges that emerge when trying to foster science/

society or science/public co-production of research and engagement within a global institution – namely, Future Earth. **Alison Mohr**, by contrast, deals with the tensions that emerge when Western energy technologies are distributed in the global South and how co-production between energy experts, social science experts and local community experts can help in this context. In both cases, openness is the *sine qua non* for such global enterprises to succeed. In the final chapter, **Stevienna de Saille** and **Paul Martin** tackle in an almost playful but deadly serious way some of the potentially problematic (or monstrous) consequences of the 'opening up' agenda written into responsible research and innovation frameworks. They do this by inspecting stories about monsters that have been told and are being told around GM foods and crops.

The part is introduced by **Barbara Prainsack** and **Sabina Leonelli**, who tease out the discursive promises and risks of using buzzwords such as 'openness' and 'responsible innovation'. They also examine the tensions explored in some of the chapters between efforts at centralisation on the one hand and opening up research and institutions to epistemic diversity on the other, as well as between inclusiveness and social justice.

Expertise

This part continues to explore issues around expertise, experts and publics. The first chapter in this part, by **Sarah Hartley** and **Adam Kokotovich**, focuses on the always hot topic of risk and risk assessment. They make the claim that public involvement in risk assessment is not reaching its full potential and argue for a new role for experts and publics, supported by a detailed analysis of a particular case study; namely, the European Food Safety Authority's public consultations. We then move from food safety to emerging diseases, in this case the emergence of a plant/tree disease: ash dieback. The chapter by **Judith Tsouvalis** finds a similar disconnect between experts and publics and a similar divorce between 'risk' assessment and public values. Both chapters make a plea for not dealing with risks from a purely expert and technoscientific perspective. **Warren Pearce** and **Brigitte Nerlich** in turn explore a particular case study, the release of the film *An Inconvenient Truth* in 2006, as an example where climate change expertise is taken out of the pages of science journals and into the public sphere, and the opportunities and problems this generates. **Sujatha Raman**, **Pru Hobson-West**, **Mimi Lam** and **Kate Millar** use a famous political speech, 'Science Matters', as an opportunity to rethink the role of engagement by minority publics in constituting the public interest around science in alliance with expertise.

This part is introduced by **Mark Brown**, who sheds light on the tensions between experts and publics by providing a historical overview of the relationship between science and democracy. He examines the legitimacy of expertise in the current political climate and points out that 'avoiding technocracy without fostering populism is a key challenge of our time'.

Faith

This part continues to explore some of the topics addressed in the previous one, dealing with expertise, experts and publics, but with a particular focus on science and religion. The chapter by **Fern Elsdon-Baker** questions the expertise of social scientists when dealing with a particular type of 'public' – namely, people who in one way or another lean towards a creationist view of life on earth. She makes a plea for researchers to not posit as a 'fact' a presumed clash between scientific and religious world views, cautioning against assuming that the latter is always a monstrous public in conflict with science, and to explore public perceptions of evolutionary science and religion without either being overshadowed by this prejudice. The second chapter, by **David Kirby** and **Amy Chambers**, is a fascinating exploration of the struggle between film-makers and religious communities over shaping public views of science, including evolution, through a history of censorship. **Chris Toumey** introduces this part, weaving together reflections on science and religion with the themes of openness, expertise and responsibility in new and unexpected ways.

The book is rounded up with two closing statements, one an afterword by **John Holmwood** and **Jan Balon** reflecting on markets, neoliberalism, populism and the demise of the public university, which is one current issue that bedevils our (academic) lives. This is followed by an epilogue by **Stephen Turner**, who weaves together all the chapters presented in this book into a coherent story, by projecting them against a much-needed historical background involving science, politics and publics.

Of course, a volume published in 2018 devoted to debates about the fractious relationship between science, policy and publics would be remiss if it were not to make more than a fleeting reference to the extraordinary year in politics that was 2016. In June, the UK voted in a referendum to withdraw from the European Union. At the time of writing, formal negotiations are under way between the British Government and their European counterparts following Prime Minister Theresa May's decision to trigger Article 50, which begins the two-year countdown to 'Brexit', on 29 March 2017. The UK referendum was followed six months later by the election of the billionaire Republican Donald Trump to the White House. The outcomes of both the British referendum and the US presidential election have profound consequences for science, which only serves to emphasise – sadly, from the point of view of many of our contributors – the importance of the questions explored in this volume in the current political climate.

We therefore end this introduction with a coda by **Alexander Thomas T. Smith**, which provides a brief snapshot (as of spring 2017) of the political landscape following these two electoral events, with a focus on some of the repercussions for science funding and policy in both countries.

Coda: reflections on the politics of openness in a new world order

Alexander Thomas T. Smith

Politically, 2016 was a convulsive, tumultuous year. On 23 June, the United Kingdom held a referendum on whether to remain a member of the European Union. On a turnout of 72%, a narrow majority – 52% – voted to 'Leave' the EU. The Conservative Prime Minister David Cameron, who had called the referendum and campaigned to 'Remain', had not anticipated defeat and announced his intention to step down the day after. Within a handful of weeks, the Conservatives elected a new leader – Theresa May – who took over as prime minister on 13 July. Although she had supported Remain in the referendum – albeit without much enthusiasm publicly – May now declared that 'Brexit means Brexit' and committed her government to triggering Article 50, which would begin the two-year negotiations, and countdown, to Britain's departure from the EU.

Around the same time, across the North Atlantic, the Republican Party was about to nominate the maverick billionaire Donald Trump as its candidate for the US presidency. Trump had seen off a wide field of ex-governors and senators in a heated campaign for the nomination and now faced the Democratic Party's presumptive nominee, former Secretary of State Hilary Clinton. Few at the time imagined the pugnacious, impatient and thin-skinned Trump would defeat a veteran Democratic Party candidate like Clinton. After Britain unexpectedly voted to leave the EU, however, some began to wonder if the unthinkable might now just become possible.

The challenge for Trump remained winning enough electoral college votes to secure the White House, in so-called battleground states where there live significant minorities whom he repeatedly offended during his chaotic campaign. A month before the November election, a recording surfaced of Trump boasting about his lecherous behaviour towards women. With his campaign seemingly in meltdown, most pollsters and political commentators expected Trump to haemorrhage support, particularly from female Republican voters. Such was the seriousness of his self-inflicted political wounds, some even thought he would resign before election day. The *New York Times* gave Clinton a 92% chance of winning.

The election, held on 8 November, proved that the unthinkable could, indeed, happen. Despite losing the popular vote to Clinton by three million, Trump picked up almost all the battleground states: Florida, Iowa, Michigan, North Carolina, Ohio, Pennsylvania and Wisconsin (Clinton held Virginia). States that were rumoured some months earlier to be in contention, including Arizona where a substantial Latino population was thought to be mobilising against Trump, remained solidly Republican.

The victories of both the Leave campaign in the British referendum on EU membership and Donald Trump in the US presidential election stunned many experts, pundits and political commentators. Both results seemed to highlight volatility in the British and US electorates that had been long speculated about, but which now finally found tangible form, as democratic facts.

Most of the contributors to this volume wrote their chapters before these two political earthquakes turned many of our assumptions about what the citizenries of our respective countries are thinking on their heads. As this book goes into production, in the spring of 2017, we are troubled by a suspicion that, already, global events risk eclipsing the intellectual concerns about which our authors have written so eloquently here. Brexit and Trump have now become synonymous with instability and uncertainty. Politics, in both countries, has been anything but normal ever since.

After 100 days in the White House, it is clear that President Trump has been keen to conduct himself much as he did as a candidate, riding the crest of a nativist–populist tide and railing against various kinds of elites, in Washington DC, the mainstream media and in the universities. Those concerned about the future of science policy, including the funding of research and teaching in both the social and natural sciences, are right to be worried. Trump has threatened to cut funding to the National Science Foundation and even abolish the National Endowment for the Humanities. A long-time sceptic of human-induced climate change, he has described global warming as a 'hoax' perpetuated by the Chinese. In a provocative move, Trump appointed Scott Pruitt administrator of the Environmental Protection Agency (EPA). Pruitt, a lawyer from Oklahoma, is on record as having rejected the scientific consensus on climate change. His appointment has been destabilising to the EPA as an institution and has raised questions about the future of climate science under Donald Trump. It also generated anxiety amongst environmentalists, activists and policymakers keen to see the USA uphold its international obligations and continue to demonstrate leadership in reducing the global economy's dependence on burning fossil fuels, fears that were confirmed on 1 June 2017 when President Trump announced the US would quit the Paris climate accord. On 6 March 2017, one friend on Facebook shared with me an anonymous message from a friend of his who works at the EPA:

> [Yeah] it's as bad as you are hearing: The entire agency is under lockdown, the website, facebook, twitter, you name it is static and can't be updated. All reports, findings, permits and studies are frozen and not to be released. No presentations or meetings with outside groups are to be scheduled. Any Press contacting us are to be directed to the Press Office which is also silenced and will give no response. All grants and contracts are frozen from the contractors working on Superfund sites to grad school students working on their thesis. We are still doing our work, writing reports, doing cancer modeling for pesticides hoping that this is temporary and we will be able to serve the public soon. But many of us are worried about an ideologically-fueled purging and if you use any federal data I advise you gather what you can now. We have been told the website is being reworked to reflect the new administration's policy. … [You] all pay for the government and you should know what's going on. I am posting this as a fellow citizen and not in any sort of official capacity.

On their own, Trump's intentions vis-à-vis climate science would raise enough questions. For advocates of science who believe the latter thrives best in conditions of 'openness' – one of this book's central themes – there have also been other causes for concern.

One particular worry is that Republican lawmakers might use principles of transparency and reproducibility, which are fundamental to advancing the scientific method, to 'weaponise' science against itself, using data generated from poor-quality studies to undermine scientific credibility and consensus with a view to discrediting and defunding science programmes not to their ideological liking (see Yong, 2017). Another concern has arisen in response to calls during the early days of the Trump administration to close the border to citizens from a handful of Muslim-majority nations in the Middle East and North Africa. The White House has cited concerns about national security in its justification for promoting this policy, which was articulated alongside further calls to build a wall along the US–Mexico border and deport millions of undocumented migrants living in the country. While the so-called 'Muslim ban' has been challenged successfully in the courts, the cumulative impact of both policy proposals has been to weaken the United States' claim to be a country welcoming to immigrants and refugees. Many supporters of science worry this will discourage members of the international scientific community from collaborating with colleagues based at US institutions. Could the apparent hostility of the Trump administration towards Latino and Muslim migrants fuel a more negative impression that the USA is no longer 'open' to those seeking opportunities for collaboration, employment or study in the United States?

Similar anxieties have animated debates about the future of science and higher education in the United Kingdom following Brexit. For now, it would seem that British scholars and scientists have much to lose when the country departs the European Union. If the UK Government embraces what has been coined a 'hard' Brexit, resulting in membership of the European single market being sacrificed in favour of uncompromising controls over borders and immigration, there is a risk that UK science will be denied opportunities to access EU research funding, participate in wider exchanges such as the ERASMUS scheme (European Region Action Scheme for the Mobility of University Students), or even attract well-qualified undergraduate or postgraduate students from EU member states. British-based researchers who are not citizens of the UK but are EU nationals worry about their future beyond the conclusion of Article 50 negotiations. Anecdotal evidence suggests there has already been a negative impact on prospective collaborations between UK- and EU-based research institutions, even though Britain remains a EU member state until at least 2019 and can access European research funds during that time. Whether or not British researchers find clear answers to their many questions following the snap general election Prime Minister Theresa May called for June 2017 remains to be seen.

The uncertainties science now faces, in both the UK and the USA, underscore the importance of the issues explored in the chapters that follow. One hope of this volume is to stimulate a conversation about the relationship between science, publics and openness and it would seem, in what we might characterise as politically 'monstrous' times, the need for such a conversation has now become especially urgent. How might we, as scholars and supporters of funding for scientific research and teaching, respond in this moment of 'crisis' for the disciplines we hold dear?

There are examples from which, I would suggest, we can draw inspiration. On 22 April 2017, thousands of supporters of science marched in the United States, the UK, indeed hundreds of cities worldwide. The so-called 'March for Science' attracted considerable media attention. Such initiatives help raise awareness of the issues facing the science community as a result of the policy uncertainties that Brexit and Trump's election have unleashed.

I also take inspiration from my research subjects in Kansas. Since 2008, I have been visiting the state regularly, conducting ethnographic fieldwork on grassroots Republican Party politics (see Smith, 2013, 2015, forthcoming). My particular interest has been this: how do political moderates seek to empower themselves in the face of right-wing extremism and religious conservatism?

Kansas was the perfect field site in which to explore this question. It has been on the front line of the United States' culture wars for the last three decades. Infamously, in 2005, the State Board of Education held a series of hearings on evolutionary theory, with a view to making the case for the teaching of intelligent design in the high-school science curriculum. At the time, evangelical Christians who believed in young-earth creationism had captured the Kansas Board of Education. The hearings attracted media interest from around the globe before moderate Kansans mobilised and overturned the creationist majority on the State Board of Education in 2006.

I began this research at a time when US satirists, cultural commentators and politicos relished asking the rhetorical question that Thomas Frank (2004) popularised when he asked *What's the Matter with Kansas?* I commenced this study before the Tea Party movement came along, before the election of the socially conservative Sam Brownback as governor of Kansas, before far-right Republicans seized control of the state legislature and introduced destructive tax cuts that continue to imperil a wide range of government services, including public education. And I began this research before Donald Trump helped unify and give voice nationally to an angry nativist constituency.

But the question I sought to find answers to, in the United States' heartland, now seems more vital than ever. If Kansas has been ahead of the curve as far as US debates about the interface between science and democracy are concerned, this is where I find hope. Because on the day Donald Trump was elected president, moderate Kansans quietly turned up at their polling booths and elected a string of moderate Republicans and Democrats to the state legislature in what some are beginning to understand was an important rejection of the far-right economic and social policies that Trump appears keen to continue championing nationally. Now, with a moderate majority in Topeka, Republican and Democratic Party legislators, working together, are trying to fix the damage wrought on the state's finances after almost six years of a reckless politics. More than anything, this gives me hope – for science, for expertise, for publics that value education, for the renewal of a measured and moderate politics on both sides of the Atlantic.

If it can happen in Kansas, it can happen anywhere.[1]

This coda was written at the beginning of 2017.

---

1 National and international media organisations extensively covered both the UK referendum on EU membership and the election of Donald Trump as US president. My primary sources here are the BBC, the *Guardian* and the *New York Times*.

References

Brown, M. B. (2009). *Science in democracy: Expertise, institutions, and representation*. Cambridge, MA: MIT Press.

Chilvers, J., and Kearnes, M. (2016). Science, democracy and emergent publics. In J. Chilvers and M. Kearnes (eds), *Remaking Participation: Science, Environment and Emergent Publics* (pp. 1–28). London and New York: Routledge.

Collins, H. (2014). *Are We All Scientific Experts Now?* London: John Wiley and Sons.

Collins, H. M., and Pinch, T. (1993). *The Golem: What You Should Know about Science*. Cambridge: Cambridge University Press.

Delingpole, J. (2010). 'Postnormal-science' is perfect for climate demagogues: It isn't science at all. *Spectator*, February. Retrieved 14 December 2016 from: www.spectator.co.uk/2010/02/postnormal-science-is-perfect-for-climate-demagogues-it-isnt-science-at-all/.

Frank, T. (2004). *What's the Matter with Kansas? How Conservatives Won the Heart of America*. New York: Owl Books.

Funtowicz, S., and Ravetz, J. (1993). Science for the post-normal age. *Futures*, 25(7), 739–755.

Habermas, J. (1970). *Toward a Rational Society: Student Protest, Science, and Politics*, trans. by J. J. Shapiro. Boston, MA: Beacon Press.

Haraway, D. (1992). The promises of *monsters*: A regenerative politics for inappropriate/d others. In L. Grossberg, C. Nelson and P. A. Treichler (eds), *Cultural Studies* (pp. 295–337). New York: Routledge.

Hartley, S., Gillund, F., van Hove, L., and Wickson, F. (2016). Essential features of responsible governance of agricultural biotechnology. *PLoS Biology*, 14(5), e1002453.

Holmwood, J. (2011). *A Manifesto for the Public University*. London: Bloomsbury Academic.

Holmwood, J. (2013). Commercial enclosure: Whatever happened to open access? *Radical Philosophy*, September/October. Retrieved 14 December 2016 from: https://www.radicalphilosophy.com/commentary/commercial-enclosure.

Irwin, A., and Michael, M. (2003). *Science, Social Theory and Public Knowledge*. London: McGraw-Hill Education.

Jasanoff, S. (2002). Citizens at risk: Cultures of modernity in the US and EU. *Science as Culture*, 11(3), 363–380.

Jasanoff, S. (2003). Technologies of humility: Citizen participation in governing science. *Minerva*, 41(3), 223–244.

Jasanoff, S. (2006). Transparency in public science: Purposes, reasons, limits. *Law and Contemporary Problems*, 69(3), 21–45.

Law, J. (1991). Monsters, machines and sociotechnical relations. In J. Law (ed.), *A Sociology of Monsters: Essays on Power, Technology and Domination* (pp. 1–23). London: Routledge.

Mance, H. (2016). Britain has had enough of experts, says Gove. *Financial Times*, 3 June. Retrieved 5 July 2016 from: https://www.ft.com/content/3be49734-29cb-11e6-83e4-abc22d5d108c.

Otto, S. L. (2016). *The War on Science: Who's Waging It, Why It Matters, What We Can Do about It*. Minneapolis, MN: Milkweed Editions.

Owen, R., Stilgoe, J., Macnaghten, P., Gorman, M., Fisher, E., and Guston, D. H. (2013). A framework for responsible innovation. In R. Owen, J. Bessant and M. Heintz (eds), *Responsible Innovation: Managing the Responsible Emergence of Science and Innovation in Society* (pp. 27–50). London: John Wiley.

Sarewitz, D. (2004). How science makes environmental controversies worse. *Environmental Science and Policy*, 7(5), 385–403.

Smith, A. T. T. (2013). Democracy begins at home: Moderation and the promise of salvage ethnography. In A. T. T. Smith and J. Holmwood (eds), *Sociologies of Moderation: Problems of Democracy, Expertise and the Media* (Sociological Review monograph) (pp. 119–140). Oxford: Wiley Blackwell.

Smith, A. T. T. (2015). Kansas versus the creationists: Religious conflict and scientific controversy in America's heartland. In S. Brunn (ed.), *The Changing World Religion Map* (pp. 997–1011). New York: Springer Academic Press.

Smith, A. T. T. (forthcoming). *Democracy Begins at Home: Political Moderation in Red State America*. Lawrence, KS: University Press of Kansas

Smith, A. T. T., and Holmwood, J. (2013). *Sociologies of Moderation: Problems of Democracy, Expertise and the Media*. Oxford: Wiley Blackwell.

Stilgoe, J., and Burall, S. (2013). Windows or doors? Experts, publics and open policy. In R. Doubleday and J. Wilsdon (eds), *Future Directions for Scientific Advice in Whitehall* (pp. 92–99). N.p.: University of Cambridge, Centre for Science and Policy; Science Policy Research Unit (SPRU), ESRC STEPS Centre at the University of Sussex; Alliance for Useful Evidence; Institute for Government; and Sciencewise.

Turner, S. (2003). *Liberal Democracy 3.0: Civil Society in an Age of Experts*. London: Sage Publications.

University of Manchester, Institute for Science, Ethics and Innovation (2010). *Who Owns Science? The Manchester Manifesto*. Retrieved 14 December 2016 from: www.isei.manchester.ac.uk/TheManchesterManifesto.pdf.

Watermeyer, R. (2012). *Written Evidence Submitted by ESRC Centre for Economic and Social Aspects of Genomics (Cesagen) Cardiff University (PE 01)*. Commons Select Committee, Public Administration Committee. Retrieved

14 December 2016 from: www.publications.parliament.uk/pa/cm201314/cmselect/cmpubadm/75/75we02.htm.

Wynne, B. (1992). Misunderstood misunderstanding: Social identities and public uptake of science. *Public Understanding of Science*, 1(3), 281–304.

Yong, E. (2017). How the GOP could use science's reform movement against it. *Atlantic*, 5 April. Retrieved 26 April 2017 from: www.theatlantic.com/amp/article/521952/.

Part I

Transparency

1

Transparency

Benjamin Worthy

Christopher Hood and David Heald define transparency as a glass-like process allowing those outside to look in, a metaphoric and literal 'peering through the window' illustrated by the glass of the German Reichstag (Heald, 2006; Hood and Heald, 2006). The concept is synonymous with openness and is rooted in the idea of letting in light or knowledge (Bok, 1986). Transparency over the last decade has been entrenched within political discourse as a kind of universal good that is both an instrumental means to a number of positive outcomes (such as improved trust or accountability) and an end in itself (Meijer, 2013). It is, moreover, an idea that is universally supported across the political spectrum as a means of opening up institutions to public scrutiny (Birchall, 2014).

Underneath this acceptance, transparency can entail many things. Darch and Underwood describe it as an 'ideologically-determined political initiative that can be deployed to achieve a range of different agendas' (2010: 4). As these chapters show, the exact dynamics and divisions of the use of the term vary from country to country and area to area. Transparency resembles democracy itself, in that there is a general consensus on the concept but its interpretation is 'open to complexity, contradiction and numerous varieties': it is in some senses an empty signifier that can be filled by very different interpretations or emphases (Stubbs and Snell, 2014: 160). It can have numerous different aims and purposes, from monitoring by the public to hierarchical control of lesser bodies (Heald, 2012). Below are just three examples of what transparency can mean:

- **Political empowerment:** it is a highly politicised instrument of empowerment, embodying different democratic norms and values (Fenster, 2012a).
- **Policy solution:** it can be a 'dramatically satisfying answer to every crisis and question about the state' (Fenster, 2015).
- **Economic improvement:** it is a means of increasing efficiency and even wealth, connected to a 'consumer-citizen' idea of delivery and performance measurement.

On a symbolic level, transparency policy can then be used as a radical weapon of empowerment, a tool of modernisation and a means of demonstrating an institution or organisation is more ethical, more honest or more trustworthy. Such a policy can 'allow an incumbent to make credible promises of greater transparency and anti-corruption efforts to a wary public' (Berliner, 2014: 479). It also represents an 'apparently simple solution to complex problems – such as how to fight corruption, promote trust in government, support corporate social responsibility, and foster state accountability' and is an acceptable response to problems 'at moments of crisis or moral failure', a 'visible response to public disquiet [with] attractive, palliative qualities for politicians and CEOs who want to be seen to be doing rather than reflecting' (Birchall, 2014: 77).

However, beneath the symbolism of any openness policy its dominant message is fundamentally contested. Questions asking what sort of transparency is created, of whom and by whom, expose the complex politics underlying its use (Berliner, 2014). As the famous freedom-of-information (FOI) campaigner Tom Blanton said, the goal to initiate transparency is 'not a sudden conversion' but one created by the 'specific conditions of competition for political power' (quoted in Darch and Underwood, 2010: 64). There is a constant, highly politicised struggle to define what a policy of openness can and should do (Fenster, 2015; Yu and Robinson, 2012). Classic arguments for transparency are rooted in both rational choice theories of behaviour change and principled ideas of the moral rightness of openness (Birchall, 2014; Chambers, 2004; Darch and Underwood, 2010). For governments it is often imbued with a very particular, often neo-liberal, conception of state–society relations. More radical conceptions see it as a weapon that can be used against these same neo-liberal ideas and reverse the assumptions of who is being open to whom, and debate the size of the political spaces opened (or closed) by its arrival (Birchall, 2014). Julian Assange and Theresa May are both vocal supporters of transparency, but they are unlikely to agree on what it means and who it should affect. Transparency remains a 'contested political issue that masquerades as an administrative tool' (Fenster, 2012b: 449).

Transparency is in sharp contrast with the idea of secrecy. The dangers of secrecy to becoming enlightened, and the opening up of forbidden matters resonate across mythology, from Pandora's box to the story of Faust. Modern bureaucratic secrecy grew from the mysterious aura that surrounded the divine right of kings in the seventeenth century (Bok, 1986). This secrecy was justified by the demands of security or to protect sensitive decisions, and bound up in an aura designed to elicit awe and often buried in a 'rich array of ritualistic and symbolic practices' (Costas and Grey, 2014: 1425–1426). Secrecy can be secured in the shape of formal institutional rules and regulations (such as official secrecy legislation or gagging clauses), or it can be obtained informally through unofficial concealment, taboos and socialisation (such as codes of silence, or agreed and shared needs for secrecy).

The clash between openness and secrecy equates, rather too simply, to that between good and bad, and democratic and undemocratic. There has been a growing view in democracies across the world that secrecy, or too much concealment, is 'incompatible with democracy', and it continues to be associated with evil, with 'stealth and furtiveness, lying and denial' (Bok, 1986: 8). This characterisation oversimplifies a more nuanced reality, as secrecy is closely entwined with a more positive notion of privacy, while publicity can be associated with manipulation and distortion (Bok, 1986). There are also broad swathes of social and political activity where confidentiality is accepted and deemed necessary, from the work of juries to peace negotiations, and even staunch advocate of openness and transparency Jeremy Bentham qualified the power of publicity with the need to prevent injustice (Chambers, 2004).

A further difficulty is that transparency is many things at once. It is achieved through laws but also by technology and experiments. It is partly about making politicians and institutions accountable for what they do – sometimes over minor matters and sometimes over big ones. Yet it can be, and often is, a practical tool to help people in their everyday lives. For all the attention given to scandals such as MPs' expenses, FOI legislation and access to online data are most often used to help individuals and for local or 'micro-political' issues (Worthy and Hazell, 2017).

Transparency and institutions

Transparency regulations have been slowly applied across many institutions in the past two decades, from government to central banks and from local to supranational bodies. To take the example of the UK, the Freedom of Information Act 2000, which is possibly the centrepiece of Britain's transparency regime, operates in more than 100,000 public bodies from central and local government to the police, hospitals, libraries and schools. Alongside this, specific sector-based laws have allowed people to access certain data about a range of areas from medical records to estate agents' fees, while open-data policies straddle many policy areas. Transparency regulations have also spread to the private sector, from procurement openness clauses in contracts, to publishing data on ownership of companies and multinational tax reporting.

Thus, the boundaries surrounding transparency are constantly moving. Disclosures using formal routes sit alongside leaks, semi-authorised disclosures and plants, innovations, and radical actions like WikiLeaks or the Panama papers that can all kick-start transparency and gradually shift the border between open and closed or legal and illegal disclosures (Pozen, 2013). There is rarely a clear distinction between how transparency is produced. For example, is it through an appeal to FOI laws, a leak or whistleblowing? As a consequence, the legitimacy of disclosure falls along a continuum, with government press releases at one end and Edward Snowden at the other. It is most often the government that delineates what it sees as the legal boundaries of openness; for example, of FOI laws or secrecy legislation. Government frames the narrative over where transparency begins and ends. The exact effect of transparency varies from institution to institution. Local government has long held open meetings and has probably dealt more successfully with demands for transparency than central government. Parliament has struggled with scandal but has altered its culture towards being more open (Worthy and Hazell, 2017). The overt assumption behind these changes is that transparency will trigger a chain of reactions across institutional domains:

- The public will be interested and use the information and data that are published.
- The public and others will act upon the data to leverage change across organisations.
- Transparency will trigger cultural and behavioural change within institutions.

Research increasingly questions and modifies each of these assumptions. There is no general ideal user of information and, while some openness initiatives generate public interest, others do not. FOI laws in the UK are well used, but the numerous open-data experiments have proved far more variable. As Roberts (2015) points out, the chain of events from asking for or accessing data to actually receiving them and levering change is long and weak. If or whether it can lever reforms depends on the context in which the information is placed and whether the instruments are available to enforce institutional or behavioural change as a result (de Fine Licht, 2014). The hope lying behind transparency that such information will influence or persuade rational, calculating voters or engaged citizens has not been borne out (Bauhr and Grimes, 2014). Users and voters hold 'deeply engrained' views about government and other institutions that are hard to dislodge. Any change coming about as a result of new information appears to be brief and subject specific (Marvel, 2016).

As to whether transparency drives cultural change, it can and has opened up diverse institutions from local government to parliament. How exactly an agent or body reacts to pressure for transparency can vary, from enthusiastic experimentation to minimal compliance and occasional outright resistance (Prat, 2006). Transparency reforms can transform a body or can be diluted, getting lost amid institutional wrangling or grinding to a halt against institutional resistance to change, as seen in the EU (Hillebrandt et al., 2014). While the evidence is slight for any perverse chilling effect, some politicians undoubtedly see transparency as a burden and symbolic of a constrained government, and have built a powerful counter-discourse around its (largely unsubstantiated) negative effects.

Transparency and the public

In the chapters in this part we can see that transparency policies can be a powerful tool for opening up bodies as they give citizens the 'capacity to penetrate ... defences and strategies' built up over centuries to preserve secrecy, and offer them the chance to create what Jeremy Bentham called a 'system of distrust' to monitor their rulers or those acting in their name (Bok, 1986: 9).

Generally, only a very small percentage of the population ever use the tools of transparency (Worthy and Hazell, 2017). There are a number of ways by which the public learns about transparency issues. First, they often appear in the media and political discourse as solutions to crises. Scandals, exposés or shocks, ranging from political corruption to financial crashes, create a demand for greater openness. Alternatively, the lack of openness is seen as the cause of the crisis (Roberts, 2012). Secondly, as Fenster (2015) points out, transparency has also 'captured the popular imagination' in narratives about whistleblowing or heroic leaks, such as the leaks on the 2009 MPs' expenses or those by Snowden in 2013. The existence of mechanisms such as FOI laws provide daily reminders in the media of the role and value of openness.

As all the chapters in this part mention, public interest in and use of the new transparency opportunities vary from case to case. There is a broad public awareness in the UK that some formal means of transparency exist, such as the FOI laws, and there is a general (if vague) support for them. In terms of leaks and 'radical transparency' such as Snowden's leaks and WikiLeaks, public opinion is unclear and shifts between different contexts. While there is a powerful supportive folklore on whistleblowing, expectations and concerns over national security can divide opinion as to the ethics and effects of these actions (Fenster, 2012a; Roberts, 2012). Some fascinating experiments indicate that the public support and are reassured by the presence of transparency mechanisms but have little desire to use them, instead preferring to rely on other citizens to operate them and unleash their benefits (see de Fine Licht, 2014; de Fine Licht et al., 2014).

Research shows that there are numerous disconnects over public opinion and transparency, and various factors that can shape its impact:

- **Context is key.** Although transparency is seen as a good thing, the battle over what it means and what its limits should be undoubtedly raises a series of competing and contradictory issues. Transparency overlaps with the ethics of leaks, privacy and national security. The view held by the public of any kind of transparency at any one time is highly context dependent. A leaker of classified information like Snowden may be viewed very differently than the anonymous leaker of MPs' expenses.

- **Flawed assumptions.** The idea underlying transparency – namely, that information empowers rational, calculating citizens – is misplaced, though politicians continue to press it. All those who receive information have biases, and employ heuristic assumptions that shape their ideas and views. These views may interrupt the flow or change the meaning of information disclosed. All transparency systems and instruments are shaped by the environment and political context in which they are created (Meijer, 2013).
- **Competing visions and meanings.** The debate over transparency and its effects continues, but may further complicate discussion rather than resolve it as different sides pull against each other. Governments seek a depoliticised (or redirected) transparency focused on efficiency or improving services, while activists seek greater openness of different parts of the state (and, increasingly, the private sector). The different language used and different aims may push discussion in divergent directions.
- **Fluid borders.** The exact boundaries of transparency are constantly moving. Disclosures through leaks, semi-authorised disclosures and plants, innovations such as open data, and 'radical' actions like Wikileaks can all provide an impetus to transparency and gradually move the border between 'open' and 'closed' or 'legal' and 'illegal' (Pozen, 2013). Meaning is greatly complicated by the closing off of certain issues, not least the transparency of citizens through government surveillance, a rarely mentioned aspect of the wider transparency debate (Birchall, 2014).

The chapters

The three chapters in this part discuss these themes in different areas, looking at the opportunities and pitfalls that transparency presents across three very different institutions and systems. They show it is far from simple to understand transparency.

Stephen Curry examines the potential of open access to academic research to break down barriers and open up research and knowledge to the wider public. There have been notable steps forward from funders and also from individual online innovators. However, acceptance of the idea across academia has been patchy. So far, there is evidence of some progress but there are economic and cultural barriers to the need to publish, in some cases triggered by outright resistance to the principle of engaging more widely, as seen in the case of NASA. Demand for greater access on the part of citizen users is just as uneven. Nevertheless, the momentum provided by funders and technology means open access is here to stay.

Carmen McLeod examines one of the most sensitive connections between openness and privacy in the transparency of animal testing. This chapter maps two rather different approaches, in the UK and Switzerland, and argues that information supplied about animal experiments will not be enough to mend the historically troubled relations between animal-research science and society. On the one hand, it is argued that the public have a clear right to know about what is done in their name and, on the other, there is held to be a right to privacy and a need for protection. The case illustrates how the demands of transparency must be balanced with other competing and compelling needs and rights. At its borders, transparency interacts in complex ways with other sensitive rights including privacy, confidentiality and security.

The final chapter by Roda Madziva and Vivien Lowndes on Christian asylum seekers and the UK Border Agency (UKBA) challenges some of the assumptions about openness and the idea that clear, neutral information is the bedrock of transparency. Information is profoundly shaped by its sociopolitical context and the hidden and not-so-hidden biases or ignorance that shape our environment. Their qualitative research shows how vital basic data may be misconstrued and misunderstood, and also placed within particular institutional narratives that not only close off the fairness that such procedures bring but even create perverse outcomes.

Increased openness is frequently offered by governments or organisations as a symbol of their difference from their predecessors or competitors, or their commitment to certain values and ways of working. Yet these chapters show that while transparency can change institutions, institutions can also remake transparency in their own image.

References

Bauhr, M., and Grimes, M. (2014). Indignation or resignation: The implications of transparency for societal accountability. *Governance*, 27(2), 291–320.

Berliner, D. (2014). The political origins of transparency. *Journal of Politics*, 76(2), 479–491.

Birchall, C. (2014). Radical transparency? *Cultural Studies ↔ Critical Methodologies*, 14(1), 77–88.

Bok, S. (1986). *Secrets: Concealment and Revelation*. Oxford: Oxford University Press.

Chambers, S. (2004). Behind closed doors: Publicity, secrecy, and the quality of deliberation. *Journal of Political Philosophy*, 12(4), 389–410.

Costas, J., and Grey, C. (2014). Bringing secrecy into the open: Towards a theorization of the social processes of organizational secrecy. *Organization Studies*, 35(10), 1423–1447.

Darch, C., and Underwood, P. G. (2010). *Freedom of Information and the Developing World: The Citizen, the State and Models of Openness*. Oxford, Cambridge, New Delhi: Chandos.

de Fine Licht, J. (2014). Policy area as a potential moderator of transparency effects: An experiment. *Public Administration Review*, 74(3), 361–371.

de Fine Licht, J., Naurin, D., Esaiasson, P., and Gilljam, M. (2014). When does transparency generate legitimacy? Experimenting on a context-bound relationship. *Governance*, 27(1), 111–134.

Fenster, M. (2012a). Disclosure's effects: WikiLeaks and transparency. *Iowa Law Review*, 97, 753–780.

Fenster, M. (2012b). The transparency fix: Advocating legal rights and their alternatives in the pursuit of a visible state. *University of Pittsburgh Law Review*, 73(3), 443–503. Retrieved 15 December 2015 from: http://ssrn.com/abstract=1918154.

Fenster, M. (2015). Transparency in search of a theory. *European Journal of Social Theory*, 18(2), 150–167.

Heald, D. (2006). Transparency as an instrumental value. In C. Hood and D. Heald (eds), *Transparency: The Key to Better Governance?* (pp. 59–73). Oxford: Oxford University Press for The British Academy.

Heald, D. (2012). Why is transparency about public expenditure so elusive? *International Review of Administrative Sciences*, 78(1), 30–49.

Hillebrandt, M. Z., Curtin, D., and Meijer, A. (2014). Transparency in the EU Council of Ministers: An institutional analysis. *European Law Journal*, 20(1), 1–20.

Hood, C., and Heald, D. (eds) (2006). *Transparency: The Key to Better Governance?* Oxford: Oxford University Press for the British Academy.

Marvel, J. D. (2016). Unconscious bias in citizens' evaluations of public sector performance. *Journal of Public Administration Research and Theory*, 26(1), 143–158.

Meijer, A. (2013). Understanding the complex dynamics of transparency. *Public Administration Review*, 73(3), 429–439.

Pozen, D. (2013). The leaky Leviathan: Why the Government condemns and condones unlawful disclosures of information. *Harvard Law Review*, 127, 512–635.

Pratt, A. (2006). The more closely we are watched, the better we behave? In C. Hood and D. Heald (eds), *Transparency: The Key to Better Governance?* (pp. 91–103). Oxford: Oxford University Press for the British Academy.

Roberts, A. S. (2012). Transparency in troubled times. *Tenth World Conference of the International Ombudsman Institute, November 2012*, Suffolk University Law School Research Paper 12-35. Retrieved 14 July 2016 from: https://ssrn.com/abstract=2153986.

Roberts, A. S. (2015). Promoting fiscal openness. *GIFT: Global Initiative for Fiscal Transparency*. Retrieved 6 July 2016 from: www.fiscaltransparency.net/wp-content/themes/enfold/includes/gift_embedded/en/resource_open.php?IdToOpen=20150704111.

Stubbs, R., and Snell, R. (2014). Pluralism in FOI law reform: Comparative analysis of China, Mexico and India. *University of Tasmania Law Review*, 33(1), 141–164.

Worthy, B., and Hazell, R. (2017). Disruptive, dynamic and democratic? Ten years of FOI in the UK. *Parliamentary Affairs*, 70(1), 22–42.

Yu, H., and Robinson, D. G. (2012). The new ambiguity of 'open government'. *UCLA Law Review Discourse*, 59, 178–208. Retrieved 12 January 2016 from: http://ssrn.com/abstract=201248.

2

Open access: the beast that no-one could – or should – control?

Stephen Curry

'The main thing, it seems to me, is to remember that technology manufactures not gadgets, but social change,' declared science historian and broadcaster James Burke in a lecture given in 1985 (Burke, 2005). This was several years before the rise of the personal computer and the internet. But history's knack of repeating itself means that the words are no less true of the digital transformation of the world in the last two decades. The recasting of information into digital forms that can be replicated and transmitted instantly across the globe has changed our relationship with it in myriad ways. This poses commercial challenges in some industries – music, film and newspapers, for example – but at the same time has given rise to whole new businesses such as search engines, social networking and online retailing. It has also created opportunities for the public to access public information, which is changing the provision of government services and opening up new avenues for democratic dialogue.

The effects have been no less profound within academia, even if they have been slower to work through the system. Our relationship with research papers and data is changing because it is easier and cheaper than ever before to put these scientific outputs into the public domain. In the era of printed journals this possibility had never arisen because of the costs of production and distribution. Now that these have largely disappeared, the question is: why not make all scientific research publicly available?

However, this simple question does not have a simple answer. There remains considerable debate about the extent to which open access should be allowed to perturb the mores of scholarship and research or to breach the walls of the academy. At the core of discussions on open access, at least in policy formulations, is the idea that the public, as taxpayers, should have access to the research that they fund. Academic perspectives on open access, by contrast, tend to be more focused on the internal operations of scientific research, although there are signs that the issue is stimulating discussion within the academy on how research findings should be made accessible to the public.

The growth of open access has coincided with a shift in thinking about public involvement in science, from the deficit model of public understanding of science initiatives, which tended to see the issue as one of ordinary people's lack of knowledge, to the more balanced notion of public engagement (Stilgoe et al., 2014). This makes it tricky to identify the precise effects of open access, which is the aim of this chapter. To set the scene, I will give a brief description of the open-access movement and recent policy initiatives, and discuss their impacts on the attitudes of scientists towards the broader open-science agenda and public engagement. I then consider the effects of open access (and allied moves) on the authority and independence of science – two issues that are perturbed by the increasingly blurred boundary between the academy and the public. Lastly, I examine the sometimes surprising feedback effects on open access that arise through the collaboration of advocacy groups and citizen scientists with professional researchers.

Although it lurks mainly in the background of the public engagement arena, the topic of open access nevertheless provides a useful focus for some of the broader issues raised by the interaction of public and academic domains. It sharpens the questions of what exactly the public wants or needs in terms of access to scientific research, and what the academy is prepared to yield in return for continued public support. Although open access has the capacity to change the dynamics of engagement between the public and the academy, the realisation of this potential requires examination of the balance of power between them, and clarification of the notions of academic freedom and responsibility. The journey since the 1990s suggests that no-one is in overall control of these processes. This is perhaps inevitable, and may even be desirable in a democratic society that aspires to be more open.

What is open access and how has it been implemented as a policy?

Open access is very much an academic initiative, largely conceived as a tool for researchers. Its origins lie at the messy confluence of digital technology and open licensing for software (Eve, 2014; Suber, 2012), but a defining moment appears to be Stevan Harnad's 'subversive proposal' of 1994 (Harnad, 1994). This advocated the free electronic dissemination of research results, but was envisioned as 'applicable only to ESOTERIC ... scientific and scholarly publication' to further learned enquiry by 'fellow esoteric scientists and scholars the world over'. The 2002 statement from the Budapest Open Access Initiative (Chan et al., 2010) defined open access to the research literature as:

> its free availability on the public internet, permitting any users to read, download, copy, distribute, print, search, or link to the full texts of these articles, crawl them for indexing, pass them as data to software, or use them for any other lawful purpose, without financial, legal, or technical barriers other than those inseparable from gaining access to the internet itself.

This encapsulates a broad notion of the intended audience, listing 'scientists, scholars, teachers, students, and other curious minds', but falls short of explicitly mentioning the public.

Even so, the statement did identify open access as an 'unprecedented public good', a concept used by economists to identify commodities that are non-excludable and non-rivalrous – in other words, available to all and undiminished by their use (Stephan, 2012) – and a label that draws the notion of public value into the discussion. The public dimension has certainly featured in the formulation of government open-access policies, which have tended to enshrine the rights of the public as taxpayers. David Willetts, the former UK Minister for Universities and Science, and a strong driver of open access, trailed his thinking in a speech to the Publishers' Association in May 2012: 'As taxpayers put their money towards intellectual enquiry, they cannot be barred from then accessing it' (Willetts, 2012). He did not elaborate on what the taxpayers might do with this access, despite the fact that his interest in the issue was first stirred by his own difficulties in getting hold of research papers while writing *The Pinch*, a book on the intergenerational social contract.

In a similar vein, the 2013 White House memorandum on open access stated simply that 'citizens deserve easy access to the results of scientific research their tax dollars have paid for' (Holdren, 2013). The EU's guidelines on open access for its Horizon 2020 research programme list the same goal, but also see open access as a way to 'involve citizens and society' through 'improved transparency of the scientific process' (EU, 2014). Again, the details of how citizens might be involved are left as an exercise for the citizens. Perhaps that is because making publicly funded research accessible is just one component of a broader open-data policy landscape that is shaped by a diverse set of motivations. Return on investment remains a central preoccupation for governments, and the release of public research and data is clearly seen as a way to stimulate innovation in new products, services and markets. But there is also, in the UK at least, a desire to improve public services, and a developing recognition of the link between transparency and democratic accountability – both for government and for the governance of scientific research (Boulton et al., 2012).

The emergent concept of responsible research and innovation (RRI), which places value on public input into efforts to anticipate the risks associated with novel avenues of research, such as nanotechnology or systems biology, seems likely to be co-opted as a further justification, but appears to be too recent to have figured in open-access policy formulations (Stilgoe et al., 2013).

What are attitudes to open access among scientists?

Reaction to open access among academics has been mixed. Some have embraced it enthusiastically. Others, though sympathetic in principle, have criticised various aspects of its implementation; still others have objected to the less restrictive forms of creative commons (CC) licensing as a gross infringement of authors' rights (Allington, 2014; Mandler, 2014). There have been lively internal debates between open-access advocates, publishers, learned societies, universities, funders and representatives of different scholarly disciplines (Eve, 2014; Hochschild, 2016; Mainwaring, 2016).

While most scientists seem sensitive to the resonance of open access with the amateur ethos of sharing that still survives within the research community, scientists are busy people with many pressing preoccupations, and generally only turn to the issue once a manuscript has been accepted for publication and the question of compliance with funder policies arises.

The lack of engagement among research scientists has not been helped by the convoluted history of open-access policy development. In the UK the implementation of a new policy by Research Councils UK (RCUK) in 2012–2013 stumbled at first. The original strong preference for 'gold' open access was subsequently refined to make it clear that 'green' open-access routes were acceptable alternatives (RCUK, 2014). The terms gold and green open access have yet to sink deep roots in academic or public minds and require some clarification. Gold open access entails making the published paper immediately available via the journal, which may require payment of an article-processing charge, with obvious cost implications that were not well received, particularly at a time when public funding for research was under severe pressure. Green open access generally means that the author's peer-reviewed manuscript (not yet formatted or copy-edited by the journal) is made available in an institutional repository. The green route is free to the author – its costs hidden within repository investment and the traditional subscription model of publishing. Moreover, green open access is often subject to publisher embargoes, which constitute delays of months or years before the manuscript is released to public view.

A subsequent announcement that the Higher Education Funding Council for England (HEFCE) will require all publications to be open access to qualify as submissions to future research excellence framework (REF) exercises was agnostic on whether open access should be achieved by the gold or green route (HEFCE, 2015). Although a bold move (Eve et al., 2014), this dislocation from the thrust of the RCUK policy added to confusion among researchers. A recent review of UK open-access policy by Adam Tickell has recommended the harmonisation of RCUK and HEFCE policies in order to simplify the requirements imposed on researchers (Tickell, 2016). Good policy requires effective implementation, especially by authors and their institutions. Although one of the ultimate aims of open access is the public interest, it is possible to detect in these machinations a greater preoccupation with the interests of researchers.

Some advocates see open access primarily as a service to science, its purpose being to accelerate and enrich the processes of research by freeing up public access to primary literature. Indeed, scientists (Breckler, 2006) and humanities scholars (Osborne, 2013) frequently express their suspicion that the public has little need of open access because they would not be able to understand research papers, a view sometimes deployed in defence of the status quo (Anderson, 2015).

There is little doubt that the sophistication of the research literature, coupled with its formal, jargon-laden style, is a barrier to understanding by the proverbial man or woman in the street. But such elitist views underestimate the sophistication of some members of the public, as we shall see below, and constitute a risky attitude in a democratic society. They also discount the benefits of lay summaries, which are increasingly being offered by journals, or of mediation by science bloggers (who can range more widely in an open-access landscape).

Concerns have also been expressed in some quarters that open-access policies are an infringement of academic freedom. Such arguments tend to expand the definition of academic freedom beyond its broadly agreed provisions to protect the right of scholars to investigate and publish on topics of their choosing without fear of sanction from university employers or governments. Kyle Grayson, for example, has asserted that it should also include the right 'to place your research where you believe it will have the biggest impact on the audience that you are trying to reach' (Grayson, 2013), while Rick Anderson argues it also entails 'the right to have some say as to how, where, whether, and by whom one's work is published' (Anderson, 2015). In particular, Anderson argues that these freedoms are lost in the imposition of CC BY licences, which allow free reuse by third parties (provided that acknowledgement and reference back to the original are made), and that open-access policies requiring such licences amount to 'coercion'. The coercion of academics is discussed in journalist Richard Poynder's lengthy critique of HEFCE's new policy of admitting only open-access papers in future REF assessments (Poynder, 2015). He argues that the exclusionary and divisive nature of the REF, which assesses only a minority of university researchers and teachers and is widely viewed as punishingly bureaucratic, makes it a strange bedfellow for the egalitarian impulses of the open-access movement. Poynder's view is that open-access advocates 'made a fundamental error when they sought to co-opt government to their cause'.

There are certainly some legitimate issues to be tackled here (Kingsley, 2016). The freedoms given to users of CC-BY-licensed open-access content to create derivative works remain a fixation among some humanities scholars (Mandler, 2014), fearful of remixing that obscures attribution or the author's original intent. Against this view, Eve and Kingsley have argued that such concerns are overstated and reflect an underestimation of the protections afforded by CC licences and the ethical norms of the academy (Eve, 2015; Kingsley and Kennan, 2015).

Nevertheless, it is undeniable that open access rubs up against academic freedom, as acknowledged by Curt Rice in making the case that open access can actually bolster the privileges accorded to scholars (Rice, 2013). Academic freedom remains a contested concept that should be considered negotiable as the place of scholarship within society continues to evolve. The concerns expressed by critics have mostly reflected a preoccupation with academic freedoms and rights, but it is also necessary to consider the question of responsibilities. A counter-view to Poynder is that the linkage of open access to the REF can be seen as entirely appropriate, since both are forms of public accountability. Whether funders should have a say in how the research that they support should be published is also a legitimate question, especially if the aim is to broaden the readership so that it might stir as many minds as possible, be integrated into their thoughts and give rise to new syntheses and insights. What more could – or should – a scholar hope for? That is a question that has been addressed rather cautiously, not least because researchers and their institutions remain in the thrall of journal impact factors for their career advancement and research assessment.

Has open access changed attitudes of scientists to public engagement?

Arguably, open access may serve as a useful first point of contact for many scientists with the broader audience for their research and public-facing open-science movement. But how effectively does it expose scientists to their public duty – for many, a concept defined only by the aspirational 'plans for public engagement' sections found in grant application forms?

We should note first that the UK Government's Public Attitudes to Science survey (Castell et al., 2014) has revealed a popular demand for scientists to be more involved in discussions with the public about their research. Anecdotally, there is an increasing acknowledgement by scientists of the need to interact with various public constituencies (e.g. patient groups, environmental activists, citizen scientists), though at the same time they are wary of how to go about this. For example, should such interactions happen in academic or public forums, including social media? There is also fear of exposure to the demands of advocacy groups that refuse to play by the academic rules of engagement.

Even when the encounter with open access raises questions of public interest, penetration of the academic mindset has been limited, in part because of the complexity of issues at stake. These include questions of the reliability of open-access research literature (raised by concerns about vanity publishing in author-pays business models) and the questionable peer-review quality of so-called predatory journals, both of which potentially undermine public trust in research. It is difficult to quantify the extent of these problems, which are mitigated by the desire of serious scholars to protect their reputations. To a significant extent they predate the rise of open access (Kingsley and Kennan, 2015) and address deep systemic problems with traditional peer-review processes of scholarly publishing (Smith, 2006). Related concerns have emerged more recently over the non-reproducibility of scientific findings, through either error or fraud in the original work (Casadevall and Fang, 2012). Open access is not touted as a ready solution to such ills, though the fact that it maximises the readership of the research literature clearly enhances the capacity for post-publication detection of inaccuracies.

The question of the cost of open access is one that has also exercised academics, but more because of its perceived incursions into research budgets, particularly by the gold-favouring RCUK policy, or the demands placed on authors not in receipt of research grants. The broader question of the total cost of scholarly publishing has received less attention – though it has not been ignored entirely (Hochschild, 2016; Kirby, 2015; Mainwaring, 2016). In part, this is because academics are largely ignorant of the cost of journal subscriptions, which are normally managed on their behalf by university librarians. Although the direction of travel is away from subscription models towards a totally open-access world, the details of the transition remain obscure and mired in enduring arguments between various stakeholders. Economic modelling suggests that a fully open-access publishing system could deliver savings by creating a market where there is genuine competition for publishing services (Swan and Houghton, 2012), but these have yet to be realised.

There is no easy escape from the dysfunctional features of the market in journal subscriptions, in which journals cannot be regarded as competing products by their purchasers and market forces are distorted not only by academic ignorance of the costs but also by preoccupations with journal prestige (Schieber, 2013). As a result, relatively little thought has been given by scientists to the argument that they should be supporting university librarians in their bids to get value for money in negotiations with publishers over subscription fees. There is a pragmatic case that researchers should be made sensitive to the issue of cost, especially when there is pressure on public funding, in order to avoid signalling privileged or insular attitudes. However, the complexity and lack of predictability of the pace and extent of transition to a functioning open-access market are significant impediments in this debate.

For some scientists the argument for open access is a moral one. Mike Taylor's insistence that paywalls are immoral and that the scientists' job 'to bring new knowledge into the world' requires them to make it freely accessible is a challenge to long-standing norms of the academy (Taylor, 2013). From another perspective, while welcoming the potential of open access, Hochschild has raised moral questions about its redistributive implications, particularly for poor scholars in the global South under business models that require payment of article-processing charges (Hochschild, 2016), which have yet to be answered satisfactorily.

Nevertheless, there is a growing sense that the ground is shifting in response to public need. Recent initiatives by funders and publishers to provide free access to research on Ebola virus and Zika virus in response to serious public health crises have thrown a spotlight on the slow and restrictive practices that have come to dominate publishing (Curry, 2016). In announcing the moves to speed the release of Zika virus research, the statement of the consortium of funders and publishers led by the Wellcome Trust spoke of an imperative. It was not described as a moral imperative, but did seem to resemble one. The logical corollary to these initiatives is to ask why they should not be extended to other infectious diseases – HIV, tuberculosis and malaria infections have caused more harm than Ebola or Zika – or even to other research areas where there is a strong public interest, such as antimicrobial resistance; climate change; or secure supplies of energy, water and food? Here, the idea of open access has prised open a gateway that has the potential to be transformative. It has yet to be converted into a mainstream conduit to the public domain, but the norms of the academy and its duty to the public interest are evidently still being negotiated.

The pressures towards greater openness seem increasingly irresistible and they may increase yet further as the latest generation of scientists – which has grown up with the internet – takes its seat in the academy. Among them are some notable idealists. Neuroscientist Erin McKiernan, for example, sees access to information as a human right and has pledged to work as openly as possible (McKiernan, 2015). The Open Access Button was created by two medical students, David Carroll and Joe McArthur, as a web-browser tool to help readers who encountered publisher paywalls to access free versions of the research papers they wanted (Carroll and McArthur, 2013). Sci-Hub, a freely accessible repository of over sixty-two million research papers created by software developer and neurotechnology researcher Alexandra Elbakyan, is a more radical and controversial reaction to journal paywalls. Though the repository is clearly in breach of copyright law – at the time of writing, Sci-Hub is the subject of a legal complaint initiated by Elsevier – Elbakyan has defended it by citing the provision in the UN declaration on human rights that asserts the right of everyone 'to share in scientific advancement and its benefits' (Taylor, 2016). The moral complexities, which have sombre echoes of the case brought against open-access activist Aaron Swartz after he had downloaded several million documents from JSTOR, are beyond the scope of this chapter. However, the work of these activists highlights the perspective of young people that the present state of scholarly publishing is increasingly ill fitted to the digital world. That perspective raises important ethical and technical arguments about scholarly publishing that cannot ultimately be settled in court.

How does open access affect the authority and independence of science?

Scientists commonly see themselves as part of a self-governing community of experts, and science as a responsible, self-correcting process of knowledge generation. For this reason they defend institutions such as peer review, which provides scientific control over what is published. In the UK this perspective shores up the Haldane principle, the right of scientists (within certain strategic constraints set by government and research councils) to determine which research projects should be funded. However, this view of self-governance is coming under challenge, from government transparency, journal impact and RRI agendas; from related shifts in the responsibilities owed to public engagement; from some of the public trust issues mentioned above in relation to the reliability of the research literature; and from special-interest campaign groups (e.g. on animal research, climate change, and genetic modification).

While the boundary between academic science and the rest of the world has never been impermeable, it demarcates the sphere of authority and independence of scientists. The growth of open science and social media are making this boundary more porous, and it is worth considering as a potential locus for future interactions with open access.

For the most part the relationship between scientists and social media remains guarded. Some have embraced the openness provided by new democratising channels of communication, but many continue to sneer at blogs, even those written by scientists. Although there have been cases where meaningful scientific critique has appeared in blogs, the view still prevails that these are not appropriate channels for discussions between scientists. For example, following the publication by NASA researchers of a claimed discovery of bacteria that could incorporate arsenic into their DNA, the space agency refused to engage with the critique published on the blog of microbiologist Rosie Redfield because it did not feel it to be 'appropriate to debate the science using the media and bloggers' (CBC News, 2010).

However, there is a growing sense that NASA's view is out of date. The website PubPeer.com has used the tools of social media to create a platform for open discussion of the research literature. It has emerged as a prominent venue for the identification of errors in research papers and, on occasion, instances of scientific fraud. A controversial feature of the platform is that commenters may remain anonymous (or unregistered) and their academic credentials unknown – but the scientists behind the site assert that the quality of comments from registered and unregistered users is indistinguishable (PubPeer, 2015).

The organic – some might say unregulated – growth of PubPeer reflects the enabling power of the internet and is diagnostic of unmet need in the publishing system. However, although the identity of many of its users is unknown, their comments and criticisms still largely reflect the same internal debates among researchers about quality control in the published literature. Different challenges arise when the research being discussed touches on matters of public interest or concern. Although there is a recognition that transparency is the key to developing and maintaining public trust (Boulton et al., 2012; Stilgoe et al., 2013), and that scientists have a duty to respond intelligently when confronted by challenges to their research, the upsurge in such challenges engendered by social media can pose severe difficulties. As Lewandowsky and Bishop have pointed out, 'openness can be exploited by opponents who are keen to stall inconvenient research' because campaigners may not be 'committed to informed debate' (Lewandowsky and Bishop, 2016). These may be difficult debates for the scientific community, but they are important and inevitable in a democratic society – and need to be conducted with some care (Pearce et al., 2016).

Open access has not yet assumed a prominent role in these interactions. However, they seem likely to become more frequent in an open-access world and should, it is hoped, also be better informed. Moreover, a general disposition towards openness is a core part of building trust through transparency. Experiments in open-access journals with open peer review (e.g. F1000Research, Atmospheric Chemistry and Physics) further increase the transparency of the scientific enterprise, as well as helping to mitigate some of the worst effects of anonymous peer reviews.

What is the impact of open access on the capacity of different publics to engage with science?

As noted above, open access was not primarily conceived as a service to the general public or as a driver of public engagement. Policy statements may have nodded in this direction by mentioning the rights of taxpayers to access the work they have funded, but there is a degree of blindness here because, of course, not all citizens are taxpayers.

Nevertheless, there is a broad array of public audiences for the research literature. This includes politicians and civil servants, policy researchers, media, non-governmental organisations, large and small businesses, independent scholars, graduates of various disciplines, patients' advocacy groups, and citizen scientists. The impact of open access on these groups has not yet been investigated systematically and is hard to quantify. It seems likely to be relatively minor, given that only a minority of research articles is reckoned to be open access: current estimates are that around 20%–30% of the research literature is freely available through journals or repositories, though the growth trend is upward (Laakso and Björk, 2012; Research Information Network, 2015).

That said, there are various citizens' groups that, for different reasons, want not just increased access to read the research literature but also to be able to make their own contributions to it. These include advocacy groups, particularly around healthcare and environmental issues (e.g. pollution, biodiversity), as well as the citizen-science movement. Such groups predate the internet (and open access). In the 1960s and 1970s the British Society for Social Responsibility in Science and Science for the People, for example, formed around concerns about weapons research and environmental pollution, but activists and advocates have been greatly stimulated by the organisational power of the web and the general increase in access to information that it affords. For example, a 2007 report concluded that health professionals have both underestimated the ability of patients to access and provide useful online resources and overestimated the hazards of imperfect online information (Ferguson, 2007). The threat of a phenomenon that was initially seen as a challenge to paternalistic medical practice is dissolving amid growing recognition that informed patients are valuable partners in managing healthcare.

The particular benefits of open access in this space appear patchy and uneven, perhaps owing to the relative novelty of citizen science and the still-limited extent of research that is published in this form. But there are initiatives to overcome this. PatientPower campaigns for greater access, as well as providing other sources of information, while PatientInform is an initiative run jointly by publishers, medical societies and health professionals to enable access to the research literature to member organisations (though not directly to members of the public). The demand for access is widespread – in 2006, 80% of internet users were reported to have searched online for information on at least one of sixteen different medical conditions (though demand for access to the primary research literature will only be a fraction of this). This type of search activity is most prevalent among younger people who have grown up with the internet, and seems likely to increase as they reach middle and old age.

Just as interesting is the growing involvement of patient groups in medical research, which has led to innovations that are likely to increase awareness of the potential of open access. A striking recent example springs from work on the rare genetic disorder, N-glycanase 1 deficiency (known as NGLY1). The condition was identified after Cristina and Matt Might linked up with genetics researchers in the search for the underlying cause of their young son's problematic physical and mental development. Genome sequencing identified previously unknown mutations in the N-glycanase 1 gene, and triggered the search for other patients. Thus far the case follows a pattern of parental advocacy that is familiar from Hollywood movies such as *Extraordinary Measures* or *Lorenzo's Oil*, but the interesting twist here – which is an important signal of the dynamism of patient–researcher interactions – is that the push to develop a treatment for NGLY1 has kick-started a citizen-science project (Mark2Cure) to text mine the research literature. In its first publication (made available as an open-access preprint on the bioRxiv), the project has shown that groups of citizens can identify and link keywords within the biomedical literature as accurately as a researcher with PhD-level training (Tsueng et al., 2016).

The Mark2Cure study does not have any citizen scientists as authors, but this is being normalised as an appropriate role. The open-access *British Medical Journal* 'welcomes studies that were led or coauthored by patients', while the health and social care journal *Research*

Involvement and Engagement (also open access) has a patient advocate, Richard Stephens, as a co-editor-in-chief. Beyond patient groups, citizen-scientist authors can readily be found in the literature on environmental pollution (Davis and Murphy, 2015; Padró-Martínez et al., 2015).

Similar developments – and challenges to traditional authority – are detectable across the whole spectrum of citizen-science projects, even in those areas where interest is driven by curiosity rather than personal need. Citizen-scientist projects vary enormously in scope, format and level of engagement between lay people and professional researchers (Shirk et al., 2012; Silvertown, 2009).

Attitudes to and experiences of open access vary within the citizen-science movement. Anecdotally, project organisers from the ranks of academia have reported sporadic demands for research papers that are usually satisfied on an ad hoc basis by distributing electronic versions accessed through university library subscriptions. Nevertheless, there is sensitivity to the issue. Robyn Bailey, who leads the ornithological NestWatch project at Cornell University, told me in an email that she was pleased to have been able to publish a paper co-authored with citizen scientist Gerald Clark in the open-access journal *PeerJ* (Bailey and Clark, 2014), recognising the need to share the results with all participants in the project. But she also acknowledged the pressure on academics to publish in high-impact journals, which can dramatically increase the costs if immediate access is desired.

These factors are recognised by other citizen-science projects but, although there is widespread understanding of the need to ensure that the results of such projects are made available to participants as part of a positive-feedback loop, open-access publication appears to be a relatively unusual avenue for doing so. Newsletters and blog posts serve as alternative means of communication that have the advantage of being more digestible, though for many rare diseases there are few secondary resources, and affected communities have no choice but to look at the primary literature. The Zooniverse, a diverse collection of projects, is unusual in having a clear policy requiring results to be published in open-access venues.

Citizen science is a dynamic and innovative area. Demand for access to the wider research literature seems likely to increase as the more-engaged participants seek to better understand the science behind their projects. Given the increasing sophistication of the contributions made by citizen scientists, it also seems appropriate to ensure that papers arising from their projects are made available to the whole community by open access. A recent open-access paper from the EteRNA project – an online game designed to search for improved methods for predicting the fold of RNA (ribonucleic acid) sequences – has three gamers – Jeff Anderson-Lee, Eli Fisker and Mathew Zada – as co-authors (Anderson-Lee et al., 2016). This arose because Rhiju Das, the project leader at Stanford University, noticed that Anderson-Lee and Fisker had independently compiled extensive documentation on their approach to the RNA-folding problem, and he encouraged them to write it up. EteRNA has an informal open-access policy that is about to be written into the end-user license agreement. 'It just seems like the right thing to do,' Das told me (personal communication, email, 24 February 2016).

The positive-feedback effects of open access on citizen science are important, not just for recognising citizen scientists' contributions and enhancing their knowledge and skills, but also as a way of making professional scientists more aware of the high-level capabilities of their citizen counterparts. The wider impacts of citizen science are difficult to assess, but it is an activity that could further increase the porosity of the walls of academia in ways that could have other societal benefits; for example, enhancing citizen participation in discussions around RRI.

Concluding remarks

Open access appears to fit naturally with the goal of making science public, but its particular contributions can be difficult to discern. The picture presented in this chapter is unfinished because the forces at play have yet to reach any kind of equilibrium. Though the pace of change may not be fast enough for its most enthusiastic supporters, there has been an indisputable rise of open access as a result of the advocacy of academics and the policy initiatives of governments and funding agencies. Awareness of the challenge to traditional modes of scholarly publishing is widespread within the academy, which appears to be sympathetic to open access in principle, even if the various requirements of policy implementation are not universally welcomed. Signs that it may encourage scientists to be more outward facing are emerging, but they are hard to separate from more general moves to open up the academy.

On the side of the public – or publics – levels of awareness and use are more limited. In certain quarters open access is seen as very important, but it is also just one form of research information that is available to citizens on the internet. That said, it is important to recognise that intermediaries to information such as journalists, bloggers or advocacy groups, also stand to benefit from increased open access.

The idea of open access as a journey has become something of a cliché, at least in the UK, but it retains a kernel of truth. Although the direction of travel is upwards from a relatively low baseline, the trajectory remains prone to deviation. Few would have predicted the present destination at the outset of the 1990s. Just as the diffuse boundaries between disciplines are reckoned to define a territory of creative interaction, the public–academy boundary that accompanies open access appears to be fertile ground. This is not just true of technical innovation and challenges to customs and practice in the academy. There are signs too, among academics, new publishers and citizen scientists, that it can bring new life.

As the primary producer and consumer of the research literature, the academy remains in overall control. But there are pressures from above and below for open access as part of the open-science agenda that offers the benefit of greater integration and mutual understanding between scientists and society. There are risks here, particularly in contentious areas of research that attract attention from combative campaigners, but few would contend that these can be mitigated by restricting access to the research literature. Public dialogue is an essential feature of democratic societies and can only be served by measures to increase the knowledge base of that conversation.

Acknowledgements

I am very grateful to David Willets, Dorothy Griffiths, Imran Khan and Jack Stilgoe for their insightful discussions; to Katherine Mathieson, Roland Jackson, Lydia Nicholas, Steve Royle, Robyn Bailey, Betsy

Carlson, Jason Holmberg, Chris Santos-Lang, Lea Shanley, Yonit Yogev, Emei Ma, David Sittenfeld, Rhiju Das, Chris Lintott, Grant Miller, David Slawson, David Baker, Caren Cooper, Harold Johnson, Sara Riggare, Tania Tirraoro, Mariana Campos and Ginger Tsueng for their helpful responses to email enquires; to Amy and 'OA' for comments via my blog; and to Bev Acreman, Laurence Cox, Richard Fisher, David Mainwaring and Richard Poynder for criticism of a preprint version of this chapter.

References

Allington, D. (2014). Choices, choices in the UK's two-tier scholarly publishing system: Open access and creative commons licences for funded and unfunded research. *Daniel Allington.net*. Retrieved 4 March 2016 from: www.danielallington.net/2014/08/choices-open-access-creative-commons-funded-unfunded-research/.

Anderson, R. (2015). Open access and academic freedom. *Inside Higher Ed*. Retrieved 4 March 2016 from: www.insidehighered.com/views/2015/12/15/mandatory-open-access-publishing-can-impair-academic-freedom-essay.

Anderson-Lee, J., Fisker, E., Kosaraju, V., Wu, M., Kong, J., Lee, J., Lee, M., et al. (2016). Principles for predicting RNA secondary structure design difficulty. *Journal of Molecular Biology*, 428, 748–757.

Bailey, R. L., and Clark, G. E. (2014). Occurrence of twin embryos in the eastern bluebird. *PeerJ*, 2, e273. Retrieved 25 November 2016 from https://doi.org/10.7717/peerj.273.

Boulton, G., Campbell, P., Collins, B., Elias, P., Hall, W., Laurie, G., O'Neill, O., et al. (2012). *Science as an Open Enterprise*. London: Royal Society.

Breckler, S. (2006). Open access and public understanding. American Psychological Association, *Psychological Science Agenda*. Retrieved 4 March 2016 from: www.apa.org/science/about/psa/2006/04/ed-column.aspx.

Burke, J. (2005). The legacy of science. In J. Burke, J. Bergman and I. Asimov (eds), *The Impact of Science on Society* (pp. 3–32). Honolulu: University Press of the Pacific.

Carroll, D., and McArthur, J. (2013). *Open Access Button*. Retrieved 4 March 2016 from: https://openaccessbutton.org.

Casadevall, A., and Fang, F. C. (2012). Reforming science: Methodological and cultural reforms. *Infection and Immunity*, 80, 891–896.

Castell, S., Charlton, A., Clemence, M., Pettigrew, N., Pope, S., Quigley, A., Shah, J. N., et al. (2014). *Public Attitudes to Science*. London: Department for Business, Innovation and Skills.

CBC News (2010). NASA's arsenic microbe science slammed. *CBC News, Technology and Science*, 6 December. Retrieved 23 July 2017 from: www.cbc.ca/news/technology/nasa-s-arsenic-microbe-science-slammed-1.909147?ref=rss#ixzz17P6UT100.

Chan, L., Cuplinskas, D., Eisen, M., Friend, F., Genova, Y., Guédon, J.-P., Hagemann, M., et al. (2010). Read the Budapest open access initiative. *Budapest Open Access Initiative*. Retrieved 4 March 2016 from: www.budapestopenaccessinitiative.org/read.

Curry, S. (2016). Zika virus initiative reveals deeper malady in scientific publishing. *Guardian*, 16 February. Retrieved 4 March 2016 from: www.theguardian.com/science/occams-corner/2016/feb/16/zika-virus-scientific-publishing-malady.

Davis, W., and Murphy, A. G. (2015). Plastic in surface waters of the Inside Passage and beaches of the Salish Sea in Washington State. *Marine Pollution Bulletin*, 97, 169–177.

EU (2014). Guidelines on open access to scientific publication and research data in Horizon 2020. *OpenAIRE*. Retrieved 4 March 2016 from: www.openaire.eu/guidelines-on-open-access-to-scientific-publications-and-research-data-in-horizon-2020.

Eve, M. P. (2014). *Open Access and the Humanities*. Cambridge: Cambridge University Press.

Eve, M. P. (2015). Researchers are altering their methods because of uncertainty over creative commons licenses. *Martineve.com*. Retrieved 4 March 2016 from: www.martineve.com/2015/08/22/researchers-are-altering-their-methods-because-of-uncertainty-over-creative-commons-licenses/.

Eve, M. P., Curry, S., and Swan, A. (2014). Open access: Are effective measures to put UK research online under threat? *Guardian*, 28 July. Retrieved 4 March 2016 from: www.theguardian.com/science/occams-corner/2014/jul/28/open-access-effective-measures-threat.

Ferguson, T. (2007). E-patients: How they can help us heal healthcare. Retrieved 4 March 2016 from: http://e-patients.net/e-Patients_White_Paper.pdf. [Now available at: http://participatorymedicine.org/e-Patient_White_Paper_with_Afterword.pdf.]

Grayson, K. (2013). Open access requirements will erode academic freedom by catalysing intensive forms of institutional managerialism. *LSE Impact Blog*, 9 May. Retrieved 4 March 2016 from: http://blogs.lse.ac.uk/impactofsocialsciences/2013/05/09/why-uk-open-access-threatens-academic-freedom/.

Harnad, S. (1994). Publicly retrievable FTP archives for esoteric science and scholarship: A subversive proposal. *Google Groups*, 28 June. Retrieved 4 March 2016 from: https://groups.google.com/forum/#!msg/bit.listserv.vpiej-l/BoKENhK0_00/2MF9QBO9s2IJ.

HEFCE (2015). Policy for open access in the post-2014 Research Excellence Framework: Updated July 2015. *HEFCE*. Retrieved 8 December 2016 from: www.hefce.ac.uk/pubs/year/2014/201407/.

Hochschild, J. (2016). Redistributive implications of open access. *European Political Science*, 15, 168–176.

Holdren, J. (2013). Increasing public access to the results of scientific research. *We the People, The White House, President Barack Obama*. Retrieved 4 March 2016 from: https://petitions.obamawhitehouse.archives.gov/petition/require-free-access-over-internet-scientific-journal-articles-arising-taxpayer-funded.

Kingsley, D. A. (2016). Is CC-BY really a problem or are we boxing shadows? *Unlocking Research*, 3 March. Retrieved 4 March 2016 from: https://unlockingresearch.blog.lib.cam.ac.uk/?p=555.

Kingsley, D. A., and Kennan, M. A. (2015). Open access: The whipping boy for problems in scholarly publishing. *Communications of the Association for Information Systems*, 37, 329–350.

Kirby, P. (2015). Open international relations: The digital commons and the future of IR. *E-International Relations*. Retrieved 30 June 2016 from: www.e-ir.info/2015/11/16/open-international-relations-the-digital-commons-and-the-future-of-ir/.

Laakso, M., and Björk, B.-C. (2012). Anatomy of open access publishing: A study of longitudinal development and internal structure. *BMC Medicine*, 10, 1–9.

Lewandowsky, S., and Bishop, D. (2016). Research integrity: Don't let transparency damage science. *Nature*, 529, 459–461.

McKiernan, E. (2015). About. *Why Open Research?* Retrieved 4 March 2016 from: http://whyopenresearch.org/about.html.

Mainwaring, D. (2016). Open access and UK social and political science publishing. *European Political Science*, 15, 158–167.

Mandler, P. (2014). Open access: A perspective from the humanities. *Insights*, 27, 166–170.

Osborne, R. (2013). Why open access makes no sense. In N. Vincent and C. Wickham (eds), *Debating Open Access* (pp. 97–105). London: British Academy.

Padró-Martínez, L., Owusu, E., Reisner, E., Zamore, W., Simon, M., Mwamburi, M., Brown, C., et al. (2015). A randomized cross-over air filtration intervention trial for reducing cardiovascular health risks in residents of public housing near a highway. *International Journal of Environmental Research and Public Health*, 12, 7814–7838.

Pearce, W., Hartley, S., and Nerlich, B. (2016). Transparency: Issues are not that simple. *Nature*, 531, 35.

Poynder, R. (2015). Open access and the Research Excellence Framework: Strange bedfellows yoked together by HEFCE. *Open and Shut?* [blog], 18 February. Retrieved 4 March 2016 from: http://poynder.blogspot.co.uk/2015/02/open-access-and-research-excellence.html.

PubPeer (2015). Vigilant scientists. *PubPeer* [blog], 5 October. Retrieved 4 March 2016 from: http://blog.pubpeer.com/?p=200.

RCUK (2014). RCUK policy on open access. *RCUK*. Retrieved 4 March 2016 from: www.rcuk.ac.uk/research/openaccess/policy/.

Research Information Network (2015). Monitoring the transition to open access. *Universities UK*, Open Access factsheet series. Retrieved 4 March 2016 from: www.universitiesuk.ac.uk/policy-and-analysis/reports/Documents/2015/monitoring-the-transition-to-open-access.pdf.

Rice, C. (2013). Open access: Four ways it could enhance academic freedom. *Guardian, Higher Education Network Blog*, 22 April. Retrieved 4 March 2016 from: www.theguardian.com/higher-education-network/blog/2013/apr/22/open-access-academic-freedom-publishing.

Schieber, S. (2013). Why open access is better for scholarly societies. *Occasional Pamphlet* [blog], 29 January. Retrieved 5 March 2016 from: https://blogs.harvard.edu/pamphlet/2013/01/29/why-open-access-is-better-for-scholarly-societies/.

Shirk, J. L., Ballard, H. L., Wilderman, C. C., Phillips, T., Wiggins, A., Jordan, R., McCallie, E., et al. (2012). Public participation in scientific research: A framework for deliberate design. *Ecology and Society*, 17(2), 29.

Silvertown, J. (2009). A new dawn for citizen science. *Trends in Ecology and Evolution*, 24, 467–471.

Smith, R. (2006). Peer review: A flawed process at the heart of science and journals. *Journal of the Royal Society of Medicine*, 99, 178–182.

Stephan, P. (2012). *How Economics Shapes Science*. Cambridge, MA: Harvard University Press.

Stilgoe, J., Lock, S. J., and Wilsdon, J. (2014). Why should we promote public engagement with science? *Public Understanding of Science*, 23, 4–15.

Stilgoe, J., Owen, R., and MacNaghten, P. (2013). Developing a framework for responsible innovation. *Research Policy*, 42, 1568–1580.

Suber, P. (2012). *Open Access*. Cambridge, MA: MIT Press.

Swan, A., and Houghton, J. (2012). *Going for Gold? The Costs and Benefits of Gold Open Access for UK Research Institutions: Further Economic Modelling.* Report to the UK Open Access Implementation Group. Retrieved 15 June 2015 from: http://repository.jisc.ac.uk/610/.

Taylor, M. (2013). Hiding your research behind a paywall is immoral. *Guardian*, 17 January. Retrieved 4 March 2016 from: www.theguardian.com/science/blog/2013/jan/17/open-access-publishing-science-paywall-immoral.

Taylor, M. (2016). What should we think about Sci-Hub? *Svpow.com* [blog], 22 February. Retrieved 4 March 2016 from: http://svpow.com/2016/02/22/what-should-we-think-about-sci-hub/.

Tickell, A. (2016). Open access to research: Independent advice. *Gov.uk*, Department for Business, Innovation and Skills. Retrieved 4 March 2016 from: www.gov.uk/government/publications/open-access-to-research-independent-advice.

Tsueng, G., Nanis, M., Fouquier, J., Good, B. M., and Su, A. I. (2016). Citizen science for mining the biomedical literature. *bioRxiv*, 13 June. doi: http://dx.doi.org/10.1101/038083.

Willetts, D. (2012). Public access to publicly-funded research. *Gov.uk*, Department for Business, Innovation and Skills. Retrieved 4 March 2016 from: www.gov.uk/government/speeches/public-access-to-publicly-funded-research.

3

Assuaging fears of monstrousness: UK and Swiss initiatives to open up animal laboratory research

Carmen M. McLeod

> Suspicion always attaches to mystery. ... The best project prepared in darkness, would excite more alarm than the worst, undertaken under the auspices of publicity. (Bentham, 1999 [1791]: 30)

The relationship between animal laboratory research science (AR) and society has a particularly complex, contested and troubled history and is associated with secrecy and obfuscation. Various works in the literature show societal fears that scientific experimentation on animals is a monstrous activity,[1] asking: 'What kind of person would do such an experiment?' (Merriam, 2012: 127). Two recent policy initiatives – the UK Concordat on Openness on Animal Research (UKC) and the Swiss Basel Declaration (BD) – seek to open up science–society relations and AR in order to build more trust and assuage fears of monstrousness within this space. These initiatives illustrate the challenges of negotiating or restoring trust in the relationship between science and society (see Dierkes and von Grote, 2000; Jasanoff, 2004; Wynne, 2006) and the complications of implementing an open-science agenda (Levin et al., 2016). This chapter explores the complexities of trust and openness in science and society relations through a comparative analysis of recent openness initiatives in the UK and Switzerland, examining the influence of historically troubled relations between AR science and society and considering whether the provision of more

---

1 One of the most well-known fictional accounts depicting the animal researcher as a monster is H. G. Wells's (1896) *The Island of Dr Moreau*.

information and greater transparency will be enough to mend the relationship.[2]

Historically, AR has been practised outside the purview of the public (Garrett, 2012), and intense debates between the AR community and society have sorely tested their trustful relations in the past. In the UK and Switzerland accusations of betrayal can be found on both sides of the science–society divide. In the UK, the legacy of so-called 'extremist' and violent animal-rights activities from the 1990s and the 2000s continue to taint the AR relationship.[3] Scientists and scientific institutions working in AR claimed that secrecy was necessary for security reasons as they were virtually 'under siege' (Festing, cited in Shepherd, 2007: 1). In Switzerland there is a similarly troubled relationship of trust between AR and society (Michel and Kayasseh, 2011).

In both countries the public has heard accusations of cruelty used in animal laboratory experiments, and animal-rights and anti-vivisection organisations often frame secrecy as a way to conceal activities that are unpalatable to the public. In the UK, for example, a number of undercover operations by animal-rights and anti-vivisection organisations have found that some scientists and animal-research technicians were not meeting required welfare standards towards the animals in their care.[4] These exposés suggested that scientists could not be trusted to follow procedures or apply ethical practices of animal welfare in their laboratories, and also raised concerns about the adequacy of the regulatory system governing animal research. Furthermore, the UK and Switzerland are particularly pertinent cases to consider in this

---

2 The chapter draws on data collected during a thirty-month project: Animals and the Making of Scientific Knowledge. This project included semi-structured interviews with scientists and other members of the UK AR community, and a focus group with members of the Basel Declaration Society, as well as a range of documentary sources such as organisations' webpages, newsletters, committee notes and other grey literature, along with secondary data, such as media reports.
3 For a comprehensive historical overview of animal rights 'extremism', see Monaghan (1997, 2013) and Hadley (2009).
4 A summary of undercover investigations in the UK by the British Union against Vivisection is provided by Linzey et al. (2015: 58–67).

context, as both countries claim to have among the strictest regulations worldwide governing animal research.[5]

Since 2012 there has been a dramatic rise in transparency discourses from the UK and Swiss AR communities, which emphasise the importance of greater openness about the activities, goals and justifications for continuing to use animal laboratory experiments. The BD and the UKC are key policy initiatives within these transparency discourses that aim to build trust. However, a growing social-science scholarship questions the assumption that greater transparency will necessarily improve accountability and trust within governance frameworks (Hood and Heald, 2006; Meijer, 2013; Worthy, 2010), and highlights the potential tensions between secrecy and openness (e.g. Birchall, 2011; Jasanoff, 2006; Strathern, 2000).

The 'technologies of secrets', a term employed by Holmberg and Ideland (2010) in a Swedish case study, refers to the patterns that underlie the fluid and flexible boundaries of openness and secrecy. They argue that, in the Swedish context, AR openness initiatives are often carefully stage-managed so as to allow what they term 'selective openness' in order to control (and preserve) existing power relations between science and the public. McLeod and Hobson-West (2015) suggest that, in contrast, openness initiatives in the UK, at least, are allied towards 'cautious openness', potentially allowing for greater input from interested members of the public. However, their research also highlights the variation in the discursive framing of the meanings of openness and what outcomes might be expected.

For scientists and institutions, opening up animal research also comes with attendant anxieties about the dangers of being more transparent, and whether such risks will outweigh the benefits of allowing greater public access inside the laboratory. In particular, the notions of openness, trust and mistrust must be considered against the

---

5 For example, in 1985 it was claimed that 'the current Swiss law for the protection of animals is already one of the most stringent in the world' (Jean-Jacques Dreifuss (University of Geneva), cited in MacKenzie, 1985: 17). The revised Swiss animal protection law, which came into effect in 2008, is also described as 'one of the most strict worldwide' (Swissinfo.ch, 2014). Similarly, in the UK there are frequent claims that the country has 'some of the strictest [animal research] regulations in the world' (Science Media Centre, 2013).

backdrop of the troubled relationship between AR and society, which since the early 1980s has included betrayals and controversy. In the UK, fears of the past continue to haunt AR practices. This troubled history is also pertinent in the Swiss context, as is outlined below.

The troubled history of animal research and science–society relations

The Head of the UK Animals in Science Regulation Unit (ASRU) recently commented that a 'vicious circle of distrust' has developed in the AR domain (MacArthur Clark, 2015). This narrative of distrustful relations associated with AR dates from the 1970s, particularly the impact during the 1990s and 2000s of both the increase in exposés of unethical and non-compliant activities by scientists and serious instances of hard-line and violent animal-rights activism. This historical context remains a fundamental challenge to trust in contemporary debates about transparency and AR.

The UK has a long history of AR protests, including some animal-welfare and anti-vivisection organisations that have been campaigning since the end of the nineteenth century. Although most organisations have tended to engage in non-violent forms of protest, during the 1970s and 1980s a marked increase in direct action (both legal and illegal) brought animal research into the spotlight (Matfield, 2002). From 1996 many AR breeding facilities were targeted, resulting in the closure of several of the smaller companies (Monaghan, 2013). Huntington Life Sciences, the largest contract research company in Europe, then became the focus of a campaign involving attacks on company infrastructure and staff, as well as secondary targets such as banks, stockbrokers and client companies. In 2001, owing to fears about the impact of animal-rights groups on Huntington Life Sciences, the Royal Bank of Scotland refused to renew a loan to them of $20 million. This led to concern about the future of the country's bioscience-based industries, and the UK Government introduced legislation targeting illegal animal-rights activities, which included a specific police task force with powers to arrest any person found protesting outside a private residence (Matfield, 2002).

Direct-action animal-rights activities continued throughout the 2000s against other AR-related organisations, including pharmaceutical companies (Monaghan, 2013). In addition, several UK universities were also targeted. The most high-profile (and ongoing) protest involved the University of Oxford in 2004, following the University's announcement that a new biomedical sciences building would include a rehoused animal unit. A campaign was initiated by a group called SPEAK in an effort to halt the construction of the new building, with the particular concern that primates would be housed in the unit. While this campaign began as lawful protests with letter writing and non-violent demonstrations, the co-founder of SPEAK was eventually convicted in 2009 of conspiracy to commit arson.[6] In addition, the Animal Liberation Front began publishing warnings that individuals associated with the new building (including building contractors and suppliers) 'were going to get some' (Animal Liberation Front communiqué posted on their Bite Back website, cited by Monaghan, 2013: 938).

Two incidents stand out in this historical narrative for their monstrous and distressing nature, because they involved grave robbing. In 2004 a campaign of intimidation was carried out against the Hall family who owned Darley Oaks Farm (breeding guinea pigs) in Staffordshire. This campaign included sending threatening letters to employees, and then the body of the owner's mother-in-law was removed from a cemetery by four activists who were linked to the Animal Rights Militia. Her remains were not recovered until 2006 (Ward, 2005). In media coverage the animal-rights activists responsible were described as being 'worse than animals' (Wright and Pendlebury, 2004: 9). The second case of grave robbing occurred in Switzerland in 2009, when the CEO of Swiss-owned Novartis, Daniel Vasella and his family, were targeted. An urn with Vasella's mother's ashes was stolen from the family cemetery and has never been recovered. Additionally, two crosses were placed in the family plot inscribed with the names of Daniel Vasella and his wife, also depicting a fictional date of their death (Stephens, 2009).

This historical narrative of violent direct action against AR scientists and supporters has had a powerful impact on relations between the AR community, animal-rights and animal-welfare groups, and the

---

6 SPEAK is a grassroots organisation that continues to organise protests and rallies in Oxford and Cambridge on animal-related issues. See, for example, http://speakanimalliberation.blogspot.co.uk/.

wider public. However, since 2012 there has been a general sense that the more extremist and violent actions against animal researchers and animal-research institutions and industry have largely diminished. Speaking of Research, which began in the UK as a pro-AR group and now provides international AR news, articulates a narrative of fearful scientists now being able to speak up for their research:

> Until recently scientists were afraid to talk about their own research using animals, resulting in animal rights groups monopolizing the debate on animal testing – however in the last few years all this has changed. (Speaking of Research, 2015)

In the UK this decline is attributed to a number of factors, including tighter policing leading to the imprisonment of core violent perpetrators, and the amendment and introduction of new legislation (Monaghan, 2013).[7] Several initiatives have also been developed to support people who have been affected by violence from animal-rights activists. For example, in 2004 a group called Victims of Animal Rights Extremism was set up with a membership of 100 people who had suffered violence and harassment. This group lobbied the UK Government to establish legislation specifically giving harsher convictions for illegal activities linked to animal-rights 'extreme acts' (Bhattacharya, 2004).

UK and Swiss initiatives to open up animal research

Following the decline in instances of violent and illegal actions against AR researchers and institutions, the AR community has increasingly begun to point to transparency as a means to reduce continued opposition to AR, as well as a way to address misinformation. In the UK and Switzerland, in particular, significant initiatives have emerged which portray openness as the key to building greater rapport between scientists and citizens over the use of animals in research.

Two major surveys of public attitudes to animal testing made a significant contribution to the initiation of these new transparency initiatives because they suggested there was declining public support for AR. The first was a Special Eurobarometer on Science and Technology

---

[7] Although some statistics from the USA suggest that individual scientists are more likely to be targeted now, rather than institutions (see Grimm, 2014).

carried out in 2009 across six EU countries (see YouGov, 2010). This survey reported that 84% of respondents mostly agreed that new guidelines should ban all animal experiments that cause severe pain and suffering. The survey also found that 80% mostly supported the publication of all information about animal experimentation, except confidential data that would allow the names of researchers or their work places to be disclosed. The European Coalition to End Animal Experiments (ECEAE) argued the poll highlighted a gap in understanding between the AR community and the wider public:

> The outcomes of the Eurobarometer survey prove once again that there is an obvious gap between the claims of the scientific community about animal use and public opinion about the issue. (ECEAE, 2010)

The second significant contribution to the development of new transparency initiatives was a poll carried out in the UK in 2012 funded by the Department for Business, Innovation and Skills (see Ipsos MORI, 2012). The poll suggested that support for animal research had declined, along with trust in the governance of these procedures. The poll found the number of people who object to animal research of any kind had risen (to 37%), as well as those who lack trust in the regulatory system (33%), and more than half the respondents (51%) suspected there was unnecessary duplication of animal experiments. These findings were widely reported in the media with headlines linking this growing opposition to failing trust, such as this comment from the *Guardian*: 'Public opposition to the use of animals in medical research is growing and trust in both scientists and the rules governing the controversial practice is falling' (Campbell, 2012).

There was some variation, however, in how scientists and commentators interpreted the results of the poll. For example, Professor Sir John Tooke, president of the Academy of Medical Sciences, said he was concerned at the poll's results. Stephen Whitehead, chief executive of the Association of the British Pharmaceutical Industry, saw the poll as 'a wake-up call', and a need for the UK AR community to be 'more forthright about the fact that without animal research, the bio-pharmaceutical sector cannot continue to innovate new treatments'. However, Sir Mark Walport, former head of the Wellcome Trust, denied that complacency among scientists had led to falling public support. He blamed a continuing 'environment of intimidation', which, at its most extreme, constituted 'terrorism' (cited in Campbell, 2012). This variation in responses illustrates the continuing tension for the AR community in both seeking out support and trust from the wider public through greater transparency, and also fearing dangerous or 'unruly publics' (de Saille, 2015) who may put scientists or institutions in jeopardy as a result.

The Basel Declaration

The BD was the first AR transparency initiative to emerge in Europe. It was launched in 2010 by the Basel Declaration Society, a membership organisation supported by donations from the pharmaceutical industry and other institutions affiliated to AR. The BD emerged out of a life sciences conference in Basel entitled 'Research at a Crossroads', held in November of the same year. This conference involved about eighty life-science researchers from Germany, Sweden, France, the UK and Switzerland. Sessions were focused around issues associated with non-human primates, transgenic animals, and ethics and communication with the wider public (Forschung für Leben, 2010).

The BD is a one-page document with extremely ambitious goals, and is framed as the foundational ethical framework for animal research, just as the Helsinki Declaration is for human medical research:

> Like the Helsinki Declaration, which forever altered the ethical landscape of human clinical research, the aim of the Basel Declaration is to bring the scientific community together to further advance the implementation of ethical principles ... and to call for more trust, transparency and communication on the sensitive topic of animals in research. (Basel Declaration Society, 2011)

Both individuals and organisations are encouraged to sign up to the BD. It is significant that scientists were prepared to be individual signatories (rather than via an institution), because this demonstrated a deeper, more personal commitment to the openness agenda of the BD: 'That's why it's good if it's signed by individuals rather than universities ... it's a bit more commitment in a way. You do it as a person' (Basel Declaration Society scientist, focus group, April 2015).

The Basel Declaration Society is an international grassroots organisation. By signing up to the BD,[8] signatories agree to ten fundamental principles (see Basel Declaration Society, 2011). These principles cover a range of topics relating to areas such as respecting and protecting animals, choosing research questions and experimental designs carefully, and acknowledging the importance of open communication and engagement with the public.

The UK Concordat on Openness on Animal Research

The UKC was developed through a two-step process. In October 2012 a 'Declaration on Openness on Animal Research' was launched at a widely covered media event, coordinated by Understanding Animal Research, a membership organisation that promotes and supports AR interests. At this event, over forty research institutions and funders promised to adhere to the UKC that was to be developed over the following year, followed by public consultation.[9] The final version of the UKC sets out requirements for universities, industry and related organisations to be more open about the ways in which they use animals in scientific, medical or veterinary research. Signatories are required to report annually to Understanding Animal Research about the progress of these commitments. Only organisations and institutions (not individuals, as in the Swiss case) can sign up to the UKC, and they are required to make the following four commitments: (1) to be clear about when, how and why animals are used in research; (2) to enhance communications with the media and the public about research using animals; (3) to be proactive in providing opportunities for the public to find out about research using animals; and (4) to report on progress annually and share experiences (Understanding Animal Research, 2014).

As stated earlier, the UKC was initiated as a direct response from the AR community to concerns about the declining support for animal laboratory research suggested in the results of the 2012 Ipsos MORI

---

8 In 2017, the total number of signatories to the BD (both individuals and organisations) was 4,621 (Basel Declaration Society, 2017).

9 In 2017, the number of signatories to the UKC was 116 organisations (Understanding Animal Research, 2017).

poll. The press release for the 2012 Declaration highlights the expectation that public confidence in AR will be boosted through openness about both the procedural aspects of AR and promotion of the benefits of AR:

> Confidence in our research rests on the scientific community embracing an open approach and taking part in an ongoing conversation about why and how animals are used in research and the benefits of this. (Understanding Animal Research, 2012).

Key aims in the UKC and the BD

There are three key aims that cut across the UK and Swiss animal-research openness initiatives. These aims highlight the hoped-for benefits from greater transparency of AR, but also signal the continued tensions which attach to science–society relations in this arena.

Facilitating a more informed public dialogue

Both initiatives seek to provide the public with the opportunity to be more informed about AR. The BD frames dialogue with the public on animal welfare in research as involving transparency and 'fact-based communications' (Basel Declaration Society, 2011). It is also anticipated that providing more information will benefit both supporters and critics of AR. For example, a report from a meeting organised by Understanding Animal Research and the Basel Declaration Society in 2012 states:

> These [findings from the meeting] corroborate the notion that transparency and open dialogue increase understanding of both the needs of scientists and the concerns of critics in a *mutually beneficial way*. (McGrath et al., 2015: 2430; emphasis added)

Both the BD and the UKC seek to provide more information to the public in order to counter misinformation, with the ultimate goal of achieving greater public support. However, both initiatives explicitly distance the provision of such information from a straightforwardly educational approach, and encourage 'two-way inclusive discourse' (Basel Declaration Society, 2015) and allow 'people to come to their own position on this issue' (Understanding Animal Research, 2014: 5). This first aim of improving dialogue is therefore framed around society and the provision of information to members of the public. While benefits for AR are anticipated from a better informed public, both initiatives are careful to emphasise that transparency and openness are important values in their own right. In contrast, the second aim is framed around science and the benefits that can come from greater openness between animal researchers themselves.

Building solidarity and support between animal researchers

The troubled history between AR and society outlined earlier has contributed to a sense of vulnerability for many people working in animal research. Both the BD and the UKC initiatives anticipate benefits not only to the public through the provision of more information and cooperation, but also to the AR community. The BD, in particular, has a mandate to build and support an open international AR community.

Underlying the push for improved solidarity is a presumption that greater transparency will not completely eliminate controversy or the potential for future conflict with critics of AR. This was highlighted in a *Nature* article that covered the announcement of the BD. Stefan Treue, director of the German Primate Center in Göttingen, comments: 'The animal issue is never going to go away. ... We need solidarity among all researchers' (cited in Abbott, 2010: 742). Solidarity is envisaged in different ways. Firstly, it involves the provision of support in response to direct action against the AR community. An example of this occurred in 2013 after an AR facility at the University of Milan was occupied by an animal-rights group, resulting in damage and the release of animals in the unit (Abbott, 2013). The Basel Declaration Society organised a 'Call for Solidarity' and collected 5,700 signatures from BD signatories, which were then presented at a rally in support of AR in Milan (Basel Declaration Society, 2013a).

A more complex and challenging feature of solidarity is the sharing of research data between researchers. Both the UKC and the BD encourage researchers to follow the ARRIVE guidelines,[10] which aim

---

10 The Animal Research: Reporting of in vivo Experiments (ARRIVE) guidelines were developed by the UK National Centre for Replacement, Refinement and Reduction of Animals in Research in 2010. See: https://www.nc3rs.org.uk/arrive-animal-research-reporting-vivo-experiments.

to improve and maximise information published on AR and as a consequence, minimise unnecessary and repetitive studies (McGrath et al., 2015). A position statement on the importance of open access (see chapter 2) and sharing research results was also developed following a workshop organised in London by the Basel Declaration Society and Understanding Animal Research in 2013. This statement does acknowledge, however, the challenges of increased data sharing because of potential proprietary interests in the results of AR experiments (see Basel Declaration Society, 2013b), which is a tension acknowledged in the open-science agenda more broadly (see chapter 5, and Levin et al., 2016).

This second aim of the UKC and the BD – to build solidarity between the AR community and to encourage the greater sharing of information – reveals there are always limits placed on what, and with whom, information about AR is shared. In this light, the third aim, of building trust, has increased importance.

Building trust in animal research and scientists

The final key aim identified here relates to building trust in animal research and science–society relations. In the Swiss context this goal was explicitly articulated by several scientists during a focus group meeting. One participant commented that through education and dialogue it was possible to 'to take away the fears that are there, to explain what is going on – and *then the trust can be built up* and everything works much better' (Basel Declaration Society scientist, focus group, April 2015; emphasis added). However, another scientist argued that the complexity of AR was impossible to fully explain to the public and therefore trust needed to precede openness: 'If people don't trust you, you can explain as much as you like; they will not buy it' (Basel Declaration Society scientist, focus group, April 2015).

In the UK, linking transparency to trust and confidence in animal research is also unequivocally referenced in the UKC, where signatories are asked to recognise that in order to be seen as trustworthy they are under an obligation to be 'be open, transparent, and accountable' in relation to all AR activities (Understanding Animal Research, 2014). Another example in the UK reveals how a concern that falling public trust in the AR regulatory system might impact on support for animal laboratory research funding. One of the signatories to the UKC, the Association of Medical Research Charities, explains that trust in AR governance is vital for funders, 'as they need the public's trust to continue funding work to fight diseases and find better treatments' (Nebhrajani, 2014).

The three aims outlined above illustrate how the UKC and BD seek to renegotiate society–science relations and AR under the aegis of greater transparency, but there are some ongoing difficulties and challenges to this agenda.

Challenges to renegotiating trust through more openness in AR

Yeates and Reed (2015: 504) argue that while transparency in AR sounds 'apodictically good', the value of openness initiatives must always depend on how, and with whom, information is actually shared. Signatories to the UKC are required to make specific information about their use of animals publicly available. One area where this has led to quite significant changes is the provision of information on institutional websites. A survey carried out by the author in June 2013 of ten UK university websites showed that most only had generic statements about experimentation on animals, giving very little specific information about what AR was carried out within the institution. Only two websites had details about the animal species used in research or any information about procedures. In contrast, a survey of websites of ten universities that were listed as signatories to the UKC in March 2016 revealed that all now provided details of the number and types of animal species used each year in the institution.

University College London (UCL), among other universities, has publicised this new approach to openness on its institutional website. In a 2014 *Times Higher Education* article, a senior academic explains UCL's commitment to transparency and openness and the UKC (see Else, 2014). In the same article, however, a spokesperson for the British Union for the Abolition of Vivisection welcomed greater transparency but feared that the new website was merely a public-relations exercise that sought to 'sanitise the reality of what life in a laboratory is like for animals in experiments' (Bailey, cited in Else, 2014). This comment suggests that on its own the provision of more information to wider society is not enough, and increased transparency does not necessarily lead to more trust. Philosopher Onora O'Neill (2002) has observed that transparency and openness initiatives can actually have a detrimental impact on trust because of fears that the information provided is being cherry-picked. O'Neill also highlights the centrality of confidence in the individuals and institutions that provide information, and that, without this, transparency and openness will not be enough on their own to build trust.

A further missing component in these discussions about trust and transparency is that scientists are rarely encouraged to speak about their own values and how they intersect with their research on animals. Therefore, these values remain hidden, or at least the moral and ethical ambiguities inherent to AR are almost never part of the information made available. This tendency to disassociate the personal views of animal experimenters from their work was highlighted in a 1995 report on AR scientists in the USA, which suggested that the use of dispassionate language tended to reinforce an image of scientists as cold, unethical and uncaring (Rowan et al., 1995). Over two decades later there is still very limited space allowed for AR scientists to reflect upon or discuss how their values and ethical decisions relating to their research fit into the wider socio-political and economic landscape (see McLeod and Hartley, 2017). In terms of building trust in the relationship between science and society in connection to animal research, therefore, there needs to be more openness about how values and ethics are incorporated into animal-research decision making by scientists, and how they are included in the AR regulatory framework. Of course, this is challenging, given the history of conflict on AR and the potential of making scientists more vulnerable if personal details about their values are made more accessible. This observation leads to the importance of research to understand the challenges and experiences of scientists and institutions who are being asked to embrace transparency initiatives such as the BD and the UKC. Historical studies suggest that AR scientists who feel stigmatised or threatened are less likely to be comfortable with being open about their research (Arluke, 1991; Birke et al., 2007). Participants in this current research who are actively promoting and driving forward the openness agendas in the UKC and the BD have expressed a degree of frustration that some

AR scientists still require convincing that it is safe for them to be more open about their research.

Conclusion

A historical relationship of mistrust has shaped the relationship between science and society on the topic of AR, and both sides believe that monstrousness exists on the other. Rudolf Wittkower, a historian writing on the cultural history of monsters, explained that 'monsters – composite beings, half-human, half-animal – play a part in the thought and imagery of all people at all times' (cited in Gilmore, 2003: 11). Animal research crosses this composite boundary, as animals become experimental subjects for the benefit of humans (primarily), and it is easy to understand why research involving animals elicits such cultural and social discomfiture.

The UKC and the BD both emerged out of an increased concern from the animal-research and biomedical communities that the societal mandate for conducting AR was declining. The three key aims of these initiatives discussed in this chapter suggest: (1) there is a genuine commitment to providing more information and opportunities for meaningful public dialogue; (2) the promotion of solidarity within the AR community could lead to more open access to data (although this is complicated by commercial interests); and (3) in both the UK and Swiss contexts, openness and trust are being discursively constructed as interlinking motifs.

However, it is important to recognise that transparency initiatives such as the UKC and the BD are unlikely to be enough on their own to build greater trust between the AR community and wider society. There also needs to be evidence of the trustworthiness of the AR regulatory system and the accountability processes that govern it (Dodds, 2013) as well as more opportunities for animal-research scientists (safely) to reflect upon, and make more transparent, the value-based decision making that is an inextricable part of their work, in order to then have a more productive conversation with wider society. The biggest challenge to opening up AR remains how to provide these opportunities and spaces where there can truly be inclusive, co-productive and safe conversations that move beyond caricatures of monstrous scientists or publics (see also chapter 8).

Acknowledgements

The author gratefully acknowledges the following funders for support: the Fondation Brocher, Geneva, for a two-month residential scholarship in March–April 2015, where I was able to draft this paper; and the Leverhulme Trust, as part of the research programme Making Science Public: Challenges and Opportunities (RP2011-SP-013). Special thanks also to the editors, especially Sarah Hartley, for helpful comments on this chapter.

References

Abbott, A. (2010). Basel Declaration defends animal research. *Nature*, 468: 742.
Abbott, A. (2013). Animal-rights activists wreak havoc in Milan laboratory. *Nature News*, 22 April. Retrieved 21 October 2013 from: www.nature.com/news/animal-rights-activists-wreak-havoc-in-milan-laboratory-1.12847.
Arluke, A. (1991). Going into the closest with science: Information control among animal experimenters. *Journal of Contemporary Ethnography*, 20(3), 306–330.
Basel Declaration Society (2011). Basel Declaration. *Basel Declaration Society*. Retrieved 16 March 2015 from: www.basel-declaration.org/basel-declaration/.
Basel Declaration Society (2013a). Call for solidarity/rally in Milan. *Basel Declaration Society*. Retrieved 16 March 2015 from: http://en.basel-declaration.com/meetings/organizing-committee/call-for-solidarity-rally-in-milan/.
Basel Declaration Society (2013b). *Open Access to Maximize the Value of Animal Research*. Basel: Basel Declaration Society. Retrieved 16 March 2015 from: www.basel-declaration.org/basel-declaration-en/assets/File/Workshop2_Open_Access_%20FINAL.pdf.
Basel Declaration Society (2015). *Engaging Positively and Proactively*. Basel: Basel Declaration Society. Retrieved 22 August 2016 from: www.basel-declaration.org/basel-declaration-en/assets/File/151001_Engaging%20positively%20and%20proactively%20-%20FINAL%20-%2020%20Oct%2015(1).pdf.
Basel Declaration Society (2017). *Signatures and Statistics*. Basel: Basel Declaration Society. Retrieved 17 July 2017 from www.basel-declaration.org/basel-declaration/statistics/.
Bentham, J. (1999 [1791]). Of publicity. In *The Collected Works of Jeremy Bentham: Political Tactics*, ed. by M. James, C. Blamires and C. Pease-Watkin (pp. 29–43). Oxford: Oxford University Press.

Bhattacharya, S. (2004). Scientists demand law against animal rights extremism. *New Scientist*, Daily News, 22 April. Retrieved 17 April 2014 from: www.newscientist.com/article/dn4913-scientists-demand-law-against-animal-rights-extremism/.

Birchall, C. (2011). Introduction to 'secrecy and transparency'. *Theory, Culture and Society*, 28(7–8), 7–25.

Birke, L., Arluke, A., and Michael, M. (2007). *The Sacrifice: How Scientific Experiments Transform Animals and People*. West Lafayette, IN: Purdue University Press.

Campbell, D. (2012). Public opposition to animal testing grows. *Guardian*, 19 October. Retrieved 17 April 2014 from: www.theguardian.com/science/2012/oct/19/public-opposition-animal-testing.

de Saille, S. (2015). Dis-inviting the unruly public. *Science as Culture*, 24(1), 99–107.

Dierkes, M., and von Grote, C. (eds) (2000). *Between Understanding and Trust: The Public, Science and Technology*. Amsterdam: Harwood.

Dodds, S. (2013). Trust, accountability and participation. In K. O'Doherty and E. Einsiedel (eds), *Public Engagement and Emerging Technologies*. Vancouver and Toronto: UBC Press.

ECEAE (2010). Eurobarometer survey shows public concern on animal testing. *European Coalition to End Animal Experiments*. Retrieved 8 January 2014 from: www.eceae.org/no/category/watching-brief/76/eurobarometer-survey-shows-public-concern-on-animal-testing.

Else, H. (2014). UCL launches website on animal research. *Times Higher Education*, 14 December. Retrieved 22 August 2016 from: www.timeshighereducation.com/news/ucl-launches-website-on-animal-research/2017510.article.

Forschung für Leben (2010). Research at a Crossroads? – Agenda. *Basel Declaration Society*. Retrieved 17 March 2015 from: www.basel-declaration.org/basel-declaration-en/assets/File/crossroad_programm.pdf.

Garrett, J. R. (2012). The ethics of animal research: An overview of the debate. In J. R. Garrett (ed.), *The Ethics of Animal Research: Exploring the Controversy*. Cambridge, MA, and London: MIT Press.

Gilmore, D. D. (2003). *Monsters: Evil Beings, Mythical Beasts, and All Manner of Imaginary Terrors*. Philadelphia, PA: University of Pennsylvania Press.

Grimm, D. (2014). Animal extremists increasingly targeting individuals. *Science News*, 12 March. Retrieved 28 February 2016 from: www.sciencemag.org/news/2014/03/animal-rights-extremists-increasingly-targeting-individuals.

Hadley, J. (2009). Animal rights extremism and the terrorism question. *Journal of Social Philosophy*, 40, 363–378.

Holmberg, T., and Ideland, M. (2010). Secrets and lies: 'Selective openness' in the apparatus of animal experimentation. *Public Understanding of Science*, 21(3), 354–368.

Hood, C., and Heald, D. (eds) (2006). *Transparency: The Key to Better Governance?* Oxford: Oxford University Press for the British Academy.

Ipsos MORI (2012). Views on the use of animals in scientific research. *Ipsos*. Retrieved 25 February 2013 from: www.ipsos-mori.com/researchpublications/publications/1512/Views-on-the-use-of-animals-in-scientific-research.aspx. [Now available at: https://www.ipsos.com/ipsos-mori/en-uk/views-use-animals-scientific-research.]

Jasanoff, S. (2004). Ordering knowledge, ordering society. In S. Jasanoff (ed.), *States of Knowledge: The Co-Production of Science and Social Order* (pp. 13–45). London: Routledge.

Jasanoff, S. (2006). Transparency in public science: Purposes, reasons, limits. *Law and Contemporary Problems*, 69, 21–45.

Levin, N., Leonelli, S., Weckowska, D., Castle, D., and Dupré, J. (2016). How do scientists define openness? Exploring the relationship between open science policies and research practice. *Bulletin of Science, Technology and Society*, 36(2), 128–141.

Linzey, A., Linzey, C., and Peggs, K. (2015). *Normalising the Unthinkable: The Ethics of Using Animals in Research*. Oxford: Oxford Centre for Animal Ethics.

MacArthur Clark, J. (2015). Science and society: Why the 3Rs matter. *Plenary Address to CCAC National Workshop – Enhancing Ethics and Welfare, 30th May 2015, Montreal, Canada*. Retrieved 14 August 2015 from: www.ccac.ca/Documents/National_Workshops/2015/presentations/Science_and_Society-Why_the_3Rs_Matter-Judy_MacArthur-Clark.pdf.

McGrath, J. C., McLachlan, E. M., and Zeller, R. (2015). Transparency in research involving animals: The Basel Declaration and new principles for reporting research in BJP manuscripts. *British Journal of Pharmacology*, 172(10), 2427–2432.

Mackenzie, D. (1985). Swiss to vote on ban on vivisection. *New Scientist*, 28 November, 17.

McLeod, C., and Hartley, S. (2017). Responsibility and laboratory animal research governance. *Science, Technology and Human Values*, doi: 10.1177/0162243917727866.

McLeod, C., and Hobson-West, P. (2015). Opening up animal research and science–society relations? A thematic analysis of transparency discourses in the United Kingdom. *Public Understanding of Science*, 25(7), 791–806 [epub ahead of print], doi: 10.1177/0963662515586320.

Matfield, M. (2002). Animal experimentation: The continuing debate. *Nature Reviews: Drug Discovery*, 1, 149–152.

Meijer, A. (2013). Understanding the complex dynamics of transparency. *Public Administration Review*, 73(3), 429–439.

Merriam, G. (2012). Virtue, vice and vivisection. In J. R. Garrett (ed.), *The Ethics of Animal Research: Exploring the Controversy*. Cambridge, MA, and London: MIT.

Michel, M., and Kayasseh, E. S. (2011). The legal situation of animals in Switzerland: Two steps forward, one step back – many steps to go. *Journal of Animal Law*, 7, 1–42.

Monaghan, R. (1997). Animal rights and violent protest. *Terrorism and Political Violence*, 9(4), 106–116.

Monaghan, R. (2013). Not quite terrorism: Animal rights extremism in the United Kingdom. *Studies in Conflict and Terrorism*, 36(11), 933–951.

Nebhrajani, S. (2014). Animal research: Important steps forward. *Association of Medical Research Charities* [blog], 11 April. Retrieved 14 August 2015 from: www.amrc.org.uk/blog/animal-research-important-steps-forward.

O'Neill, O. (2002). Lecture 4: Trust and transparency. *Reith Lectures 2002: A Question of Trust*. BBC. Retrieved 5 May 2016 from: http://downloads.bbc.co.uk/rmhttp/radio4/transcripts/20020427_reith.pdf.

Rowan, A. N., Loew, F. M., and Weer, J. C. (1995). *The Animal Research Controversy: Protest, Process and Public Policy – an Analysis of Strategic Issues*. North Grafton, MA: Center for Animals and Public Policy, Tufts University School of Veterinary Medicine.

Science Media Centre (2013). Briefing notes on the use of animals in research. *Science Media Centre*. Retrieved 17 April 2015 from: www.sciencemediacentre.org/wp-content/uploads/2013/02/SMC-Briefing-Notes-Animal-Research.pdf.

Shepherd, J. (2007). The mice that roared. *Guardian*, 13 February, Education, 1.

Speaking of Research (2015). The UK experience. *Speaking of Research*. Retrieved 26 March 2016 from: https://speakingofresearch.com/about/the-uk-experience/.

Stephens, T. (2009). Animal rights extremists step up pressure. *SwissInfo.ch*, 5 August. Retrieved 13 March 2015 from: www.swissinfo.ch/eng/animal-rights-extremists-step-up-pressure/979654.

Strathern, M. (2000). The tyranny of transparency. *British Educational Research Journal*, 26(3), 309–321.

Swissinfo.ch (2014). Animal testing numbers down. *SwissInfo.ch*, 26 June. Retrieved 29 January 2016 from: www.swissinfo.ch/eng/experiments_animal-testing-numbers-down/40474540.

Understanding Animal Research (2012). *Declaration on Openness on Animal Research*. Retrieved 21 October 2013 from: www.understandinganimalresearch.org.uk/files/9614/1041/0310/declaration-on-openn.pdf.

Understanding Animal Research (2014). *Concordat on Openness in Animal Research in the UK*. London: Understanding Animal Research. Retrieved 23 April 2015 from: www.cam.ac.uk/files/concordat.pdf.

Understanding Animal Research (2017). *Concordat on Openness on Animal Research in the UK*. Retrieved 17 July 2017 from: http://concordatopenness.org.uk.

Ward, D. (2005). 'We give up', says family besieged by activists. *Guardian*, 24 August. Retrieved 17 March 2015 from: www.theguardian.com/uk/2005/aug/24/animalwelfare.businessofresearch.

Worthy, B. (2010). More open but not more trusted? The effect of FOI on the UK central government. *Governance*, 23(4), 561–558.

Wright, S., and Pendlebury, R. (2004). Worse than animals: Extremists sink to new depths of depravity by digging up 82-year-old's grave … in the name of guinea pigs' rights. *Daily Mail*, 9 October, 9.

Wynne, B. (2006). Public engagement as a means of restoring public trust in science: Hitting the notes, but missing the music? *Community Genetics*, 9, 211–220.

Yeates, J. W., and Reed, B. (2015). Animal research through a lens: Transparency on animal research. *Journal of Medical Ethics*, 41, 504–505.

YouGov (2010). Special Eurobarometer: Science and technology report. *European Commission*. Retrieved 17 April 2015 from: http://ec.europa.eu/public_opinion/archives/ebs/ebs_340_en.pdf.

4

What counts as evidence in adjudicating asylum claims? Locating the monsters in the machine: an investigation of faith-based claims

Roda Madziva, Vivien Lowndes

Since around 2007, evidence-based policy (EBP) has emerged as a buzzword intended to signal the end of conviction-driven, ideological politics and heralding the aspirations for policymaking to be anchored in 'evidence and to deliver what works unsullied by ideology or values considerations' (Botterill and Hindmoor, 2012: 367; Clarence, 2002). The political impetus and preoccupation with activities associated with the idea of EBP are widespread. The belief that rational evidence will strengthen the basis for policymaking has been widely welcomed in many policy areas, including in contested spheres such as immigration.

As well as being an issue of profound contemporary relevance, immigration is a highly politicised field and the focus of moral and ideological contestation. Thus, evidence, assumed to speak for itself (Wesselink et al., 2014), has been called upon as a neutral arbiter in resolving perceived immigration problems and as one way of transcending ideological and humanistic conflicts (Spencer, 2011). Writing in the context of the various immigration policy crises, Boswell (2009) has shown how policymakers have often sought to find solutions to perceived problems of trust and legitimacy by turning to evidence in the form of expert knowledge. In this way, the role envisaged for evidence illustrates the trend towards openness and transparency as a way of generating renewed trust and legitimacy. Indeed, in political rhetoric, successive UK governments have routinely expressed a commitment to opening up immigration debates to allow policy and decisions to be influenced by reliable evidence rather than emotion and prejudice in order to increase transparency and build public trust in relation to immigration issues (Green, 2010).

However, more generally, the presentation of evidence as rational and neutral has raised important questions of how evidence is identified, mobilised and adjudicated in the policy process (Lowndes, 2016). Among other things, critics have shown that the perception that evidence is neutral overlooks the significance of the context in which evidence is produced. As Wesselink et al. (2014: 342) argue: 'what is policy-relevant evidence is determined by context. EBP's rhetoric looks for neutral, context-free and universally applicable evidence [which] fits badly with this reality.' Closely related to this is the observation that evidence must ultimately be interpreted, a process that many maintain involves persuasion and argument (Clarence, 2002; Kisby, 2011; Majone, 1989). As argued by Majone (1989) evidence exists only in the context of an argument; thus, it differs from data (raw material) or information (categorised data). Moreover, the way that evidence is interpreted is subject to individuals' understanding of the social world and what they consider to be important (Clarence, 2002: 5).

For instance, the assumptions of the policymaker or civil servant working in a policy context may determine what is understood as constituting evidence, the selection or prioritising of one form or indeed a specific piece of evidence over another and the interpretation of that evidence in the development of an argument (Kisby, 2011: 123). Thus, decisions often reflect not only beliefs about what works but judgements about what is feasible as well as elements of ideological faith, conventional wisdom and habit (Botterill and Hindmoor, 2012: 369). The presentation of evidence as neutral may serve to obscure the real political judgements and serendipities involved in contested policy areas like asylum. However, insights can be drawn from Pearce and Raman's (2014: 390) work on the new randomised controlled trials that interrogates the ways in which appeals to evidence are made as a way of opening up the limits of expertise. These authors have suggested what good evidence would look like, emphasising the need for evidence to fulfil at least three key principles; namely, attentiveness to plurality, diversity and institutions.

This chapter presents a critical analysis of the ways in which evidence is identified and mobilised in the asylum process. It looks at how evidence is actively constructed, embodying processes of meaning making that are underpinned by particular sets of power relations. The chapter draws on research from an extreme case – the adjudication of faith-based asylum claims – the characteristics of which enable us to locate the monsters in the machine more readily. In the spirit of this book we make use of the monster metaphor throughout the chapter to highlight issues of potential discrimination, prejudice and bias. We also identify a number of other issues inherent in the adjudication of faith-based claims and highlight the challenges to evidence-based approaches to the determination of refugee status. To better understand the limits of claims to openness and transparency via EBP, we have researched the experience of Christian asylum seekers, analysing key informants' narratives and Home Office assessment processes and policy documents.

We begin with a brief description of the research upon which this chapter is based. We then present a discussion of the lived experiences of Pakistani Christians seeking asylum, in three sections. We start with the context of the experience of reception, followed by the problems of evidencing the Christian faith and then the challenges of evidencing persecution, before we turn to our conclusions.

Research design and methodology

This chapter is based on qualitative research conducted between June and December 2015. Data were collected from forty research participants through interviews, focus groups and informal conversations as well as individual case reviews. The sample includes fifteen Pakistani Christians (five refugees and ten asylum seekers – five women and ten men), with the other twenty-five participants consisting of migrant support organisations; churches; Pakistani Christian community leaders; and professionals such as legal advisors, immigration judges, and those working in interpretation and translation. Snowball sampling and existing contacts facilitated our research access to these participants. The research aimed to gain an understanding of Christian asylum seekers' experiences of seeking asylum in the UK.

The research encounters were audio recorded and transcribed before the analysis, using thematic and conversational techniques. While the study has delivered depth, the findings cannot be taken to be representative of the experience of Pakistani Christians in the UK or indeed of Christian refugees from Muslim-majority countries more generally. Our intention was to study the experience of the particular individuals we spoke to and to draw 'analytical generalisations' (Yin, 2003) – that is, propositions that could then form the basis of research with a wider sample and in a variety of locations.

Pakistani Christian asylum seekers' arrival and UK policy context

Although there is a long history of migration from Pakistan to the UK, the population of Pakistani nationals seeking asylum in the UK became significant towards the end of the 1990s in response to the socio-political and religious repressions prevalent in their country of origin. The continual deterioration in Pakistan's human rights situation, particularly in the context of the country's infamous blasphemy laws, which foster the persecution of minority groups such as Christians, has seen the country ranked the sixth highest asylum-producing nation by the United Nations High Commissioner for Refugees (UNHCR) in 2014. Concurrently, the UK was rated among the top destinations for Pakistani asylum seekers (UNHCR, 2013). Correspondingly, more recent asylum statistics (Home Office, 2015a) have shown that Pakistani nationals constituted the second largest group (registering 2,302 cases, after Eritrea, with 2,583) of all asylum applications lodged in the UK in the year ending June 2015. However, in the absence of information on how many of these applications were lodged on the grounds of their Christian faith, Pakistani asylum seekers are regarded as a homogenous group.

We argue that the presentation of Pakistani asylum seekers as a homogenous ethnic group has the danger of masking other individualised identities such as religion or faith, which may in turn obscure the context of the reception experience.

The socio-political atmosphere of the UK is characterised by public and political discourses on growing asylum and immigration flows, ethnic and faith diversity, and their supposed link to community tensions and even terrorism (Joppke and Torpey, 2013). Particularly with regard to faith, as a presumed secular society the UK presents us with a paradox. On the one hand, successive governments have continued to show a public policy interest in faith communities that are often portrayed as providing moral leadership, social networks and access to hard-to-reach groups. Yet on the other hand, faith has become an unsettling aspect of multiculturalism, not only in the UK, but in Europe as a whole, especially in the terrorist attacks that came after September 11, 2001 (Dinham et al., 2009). More specifically, in the UK the disturbances in Bradford, Burnley and Oldham in 2001 and the subsequent 7/7 bombings in London in 2005 have led to public criticism of the concept of multiculturalism for its overly tolerant approaches to cultural difference, leading to a growth in diversity, segregated societies and the promotion of bad faith (extremism), often associated with Muslim identities (Lentin, 2011). Indeed, issues of the perceived and real problems of the integration of Muslims, and questions about accommodating Islam as a religion, are at the heart of current public policy debates, especially as the current migration crisis continues to unfold, and as Muslim identities become increasingly framed by global events (Statham and Tillie, 2016).

Moreover, the rhetoric of the perceived failure of multiculturalism, especially by political leaders across Europe (e.g. Cameron, 2011) has been juxtaposed with the claim to racelessness (Lentin, 2011). In the context of the claim that society in the UK is now a post-racial one, intolerance towards particular groups of immigrants has come to be justified on the basis of their cultural or religious incompatibility rather than their race (Statham and Tillie, 2016). Thus the political claim is that that culture or religion, rather than race, is to blame for the perceived negative aspects of diversity. When presented in this way, diversity becomes then a happy sign, a sign that racism has been overcome (Ahmed, 2007: 164).

Thus, in both political and public discourse, especially in the UK, it is increasingly claimed that it is no longer racist to talk about immigration control and that people can now have a sensible debate about immigration, where the notion of sensible involves making use of statistical evidence. Pointing out the limits to openness and transparency, Anderson argues that 'the claim to racelessness is not paralleled by a claim that immigration policies are not designed to keep out certain nationalities' (Anderson, 2013: 42), which has the danger of both promoting and concealing discriminatory practices towards particular nationalities.

Immigration controls and border inspections have given rise to perceptions that 'some bodies more than others are recognisable as [dangerous], as bodies that are out of place ... because of some trace of a dubious origin' (Ahmed, 2007: 162). Before an actual claim to asylum can be lodged, applicants must undergo an initial screening process involving checklists on their country of origin, routes of travel, their documentation of identity, and their fit in complex ethnic or religious categories.

The Pakistani Christians who participated in our research experienced their arrival and seeking asylum in the UK as putatively Islamophobic. Those we spoke to believed that the Home Office operates under the assumption that all migrants from Muslim-majority countries, by virtue of their place of origin, are Muslims. Thus the conflation of nationality and religion has led Pakistani Christian asylum seekers to believe that they are often treated as suspects – a conflation that brings Pakistani Muslims and Pakistani Christians together as one othered entity. Writing of her personal experience at the borders of New York City as a British citizen with a Muslim background, Ahmed (2007: 163) claims that 'for the body recognised as could be Muslim, which translates into could be terrorist ... the experience begins with discomfort'.

We encountered similar experiences among the Pakistani Christians we interviewed, including a male asylum seeker who explained:

> As a Christian asylum seeker from a Muslim majority country you face many obstacles in putting forward your case. The major obstacle is the place where you come from and the way you look – these are things that you can't change. Because of the way we look immigration officials don't trust us. ... They don't tell you openly that they are suspicious of you ... but through their actions and body language, you can tell that you are a suspect. The problem is you can't easily separate Christians from Muslims as we all look the same. ... I am a Christian, but when people see me they just conclude that I am a Muslim. So they think I have come to bomb their country.

Participants expressed deep concern about the equation of Pakistanis with Muslims, and in turn the equation of Muslims with terrorists

(Ahmed, 2007), showing how this multilayered stereotyping inevitably functioned to obscure their own distinctive identity as Christians from a Muslim-majority country. In the sections that follow we delve into the role of evidence in the adjudication of asylum claims.

The adjudication of asylum claims: the policy context

The UK is signatory to the 1951 Convention Relating to the Status of Refugees and its subsequent Protocol of 1967, as ratified in 1954 and 1968, respectively. According to Article 1A(2) of the 1951 Convention, an applicant for asylum must have a well-founded fear of persecution; the fear must be based on past persecution or the risk of future persecution on one or more of the specific grounds of race, religion, nationality, membership of a particular social group, or political opinion. The nature of evidence that can be provided to support such fears is a key element in the actual process of determining whether to provide asylum, as will be discussed below.

In line with the UNHCR (2004) instructions on religious persecution, the UK Home Office guidelines state that:

> persecution for reasons of religion may take various forms; for example, prohibition of membership of a religious community, prohibition of worship in private or public, prohibition of religious instruction, requirement to adhere to a religious dress code, or serious measures of discrimination imposed on persons because they practise their religion or belong to a particular religious community.

However, the Home Office goes further to note that:

> the simple holding of beliefs which are not tolerated in the country of origin will normally not be enough to substantiate a claim to refugee status. ... The issues to be decided are whether the claimant genuinely adheres to the religion to which he or she professes to belong, how that individual observes those beliefs in the private and public spheres, and whether that would place him/her at risk of persecution. (Home Office, 2015b: 28)

These expectations raise the question of the competence of immigration officials in religious matters, or the extent to which they are qualified to assess the genuineness of an individual's beliefs and the manner in which they are practised in different socio-political and religious contexts. We will return to this point later.

In the asylum process, after getting through the initial screening process, applicants still need to undergo a substantive interview in which they are interviewed by an immigration caseworker. The burden of proof lies with the applicant. This means that applicants claiming refugee status on grounds of their Christian faith are expected to establish and demonstrate their well-founded fear of persecution on the basis of their Christian identity. This involves providing a personal testimony and supporting evidence to prove that they are Christians and that they were persecuted on account of their Christian identity. Meanwhile, the UNHCR guidelines on burden and standard of proof in refugee claims (1998, para. 2) stress that, while the burden of proof lies with the claimant, decision makers are also obligated to have an objective understanding of the situation prevalent in the country of origin. It is further suggested that the actual determination of refugee status need not be certain, but must be sufficiently likely to be true. Thus, determining whether a claimant qualifies for international protection demands that decision makers judge whether they believe the applicant's evidence, or how much weight should be given to that evidence against their own understandings and interpretations of it (Thomas, 2006).

To assess the credibility of an asylum claim, immigration officials are required to consider three key criteria. The first is internal consistency, meaning that the claimant's oral testimony, written statements and any personal documents relating to the material facts of the claim should be coherent and reasonably consistent (Home Office, 2015b: 7). Secondly, external consistency is required, meaning that the claimant's testimony is expected to be consistent with the country-of-origin information or expert evidence. As stated by the Home Office (2015b: 15), 'The greater the correlation between aspects of the account and external evidence, the greater the weight caseworkers should attribute to those aspects.' The third criterion is plausibility, which is an assessment of the apparent likelihood or truthfulness of a claim 'in the context of the general country information and/or the claimant's own evidence about what happened to him or her' (Home Office, 2015b: 17).

We argue that for external evidence to be effectively used to support personal experience, it needs to properly reflect knowledge and expertise about the practical situation on the ground in the country of origin. In the Pakistani Christians' context, the reliability of the external evidence that the Home Office depends upon can be questioned. The Home Office's latest Country Information and Guidance document (February 2015) acknowledges the fact that Christians in Pakistan are generally discriminated against, distinguishing between Christian-born individuals and Christian converts, and between evangelical and non-evangelical Christians. Christian converts and evangelical Christians are perceived to be more at risk than Christian-born individuals. The same document goes on to state that 'in general the [Pakistani] government is willing and able to provide protection against such attacks' (Home Office, 2015c).

Are these distinctions between Christian-born and Christian converts and between evangelical and non-evangelical Christians useful and fair, in terms of understanding the kind of persecutions that Pakistani Christians face in their everyday lives? What sources and forms of knowledge does the Home Office rely upon? We focus on these issues in the sections below.

Evidencing Christian faith: challenges and pitfalls

From the perspective of Pakistani Christians seeking protection in the UK, proving one's faith can be a challenging exercise if the examination of it is based on biblical or doctrinal questions, which often do not seem to reflect the reality of individuals' complex identifications, denominations or practical situations in the Pakistani context. Our respondents explained how they were not only expected to know and recite certain biblical events but also to speak in certain ways that conform to Western notions of Christianity. Overall, the challenges lie in the Home Office's attempt to define people's faith technically, while at the same time assuming it must be coherent and have Western or European reference points. As one male asylum seeker told us:

> In my interview, I was asked questions like ... How do you celebrate Christmas? How do you celebrate Easter? ... and many other questions.

> I have now learnt that Christmas is a big event in this country not only for Christians but for everyone. It's regarded as a family day, no public transport because everyone is celebrating Christmas with their family. But this is not how Christmas is celebrated everywhere. In Pakistan some Christians celebrate Christmas while hiding because they don't want their neighbour to know that they are Christian. So when they ask you and you give a different answer from what they expect they say you are not a Christian.

A senior legal advisor with extensive experience of working with asylum seekers from Muslim-majority countries added that:

> the question 'How do you celebrate Christmas?' – claimants often find that very hard, because in their country of origin, Christmas is about a particular religion. But, after living here for some years, they have seen that 'Oh, Christmas is a big issue in this country.' They are even surprised by the fact that even Muslims in this country tend to give each other presents at Christmas.

The above examples show the existence of ideological perceptions about how the Christian faith ought to be manifested, which suggests that decisions to grant asylum may often be based upon a set of tacit assumptions that are not backed by evidence. In line with Pearce and Raman's (2014: 390) call for plurality, diversity and the involvement of hybrid institutions in the management of the inherent complexity of evidence, some of our participants expressed the need to open up the UK asylum system, especially by drawing knowledge and expertise from a cross-section of sources, including religious institutions. As a female vicar from the Church of England puts it:

> I think it is vital for the Home Office to consider working hand in hand with diverse churches, when it comes to faith-based claims. More notice needs to be taken of the pastors, the vicars and all those overseeing peoples' Christian journeys ... These should be respected as experts in their own right.

Such an inclusive approach could in turn help the Home Office to generate a balanced judgement of the Christian faith and the ways in which it is lived and experienced in different contexts.

Another criticism in our interviews was that, in the absence of diverse sources of evidence, the quiz-like questioning style in the assessment of the Christian faith may serve to encourage the very fraudulent claims that the UK Border Agency (UKBA) authorities fear. In the words of a legal advisor:

> Those questions favour Muslims who are recent [Christian] converts or faking to be converts because they approach the Bible like they've approached the Quran when they were little. They learn it off by heart as much as they can; it's all very fact-related.

Decision makers may seek evidence to support their own pre-existing assumptions and their very choice of evidence may be 'in itself an activity inherently lacking in neutrality' (Clarence, 2002: 5). In this way, the judges are susceptible to making incorrect decisions in two ways – either by rejecting genuine claimants or by granting refugee status to fake ones (Thomas, 2006).

Evidencing religious persecution: unveiling the monsters hiding in the machine

Existing research has shown that the UK Home Office's decision making on asylum claims suffers from a systematic and institutionalised culture of disbelief,[1] which operates alongside a parallel 'culture of denial' (Souter, 2011: 52). In our research it was common for participants to explain spontaneously why they felt they were not believed by the UKBA. However, we argue that the challenges to the credibility of the evidence offered in religious-belief cases, especially those involving accusations of blasphemy, seem to be more complex than in other refugee cases, given the uniqueness of individual cases and the need to understand the context in which they occur (see Kagan, 2010).

We observed that, for the Pakistani Christians in our study, one key area in which monsters could be hiding is in the current Home Office documents for the Pakistani country-of-origin information, particularly the distinction that is made between evangelical and non-evangelical Christians. In our research, a pattern emerged that refugee status was not granted on the grounds of religion unless it could be proved that an applicant had a religious profile in Pakistan. Yet participants' accounts of how they experienced persecution

---

1 The tendency of those evaluating applications is to start with the assumption that the applicant is not telling the truth (Home Affairs Committee, 2013: 11).

consistently revealed a stark contrast between the country-of-origin information and the actual situation on the ground in Pakistan, where Christians, regardless of their religious profile, face persecution in a country where there is limited state protection. In one of our interviews a woman asylum seeker who had been refused asylum described how practising her Christian faith in a Muslim school made her a victim of blasphemy accusations. As she explains:

> I was a teacher at a Muslim school. One day I was fasting because it was Christian Lent start date on 5th March. At break time, I was sitting in the staff room and one Muslim lady teacher ... offered me food and forced me to eat, saying, 'Take and eat.' I then said, 'No, I am fasting.' I had a big Christian magazine I was reading and she asked to see my magazine. She took the magazine and the conversation ended there because break period was over. The following day she went to the head teacher and report that I was teaching Christianity. On the Friday, this teacher's father came in school and said to me, 'You gave the magazine to my daughter ... I am giving you the chance to accept Islam.' In few days I found him waiting on the gate ... he was with a group of men ... One man punched me on the eye ... people gathered there, and the men were telling the people that she's preaching Christianity in the school.

Following the first incident, the woman and her family relocated to another place where, as she claimed, she suffered further attacks. She noted that the subsequent incident occurred in front of a local police station, but the police did not take any action. Instead, they blamed her for causing problems in the school and proceeded to file a blasphemy case against her on the instructions of her accusers. While it is clear from the claimant's account that she suffered persecution merely by virtue of her practising her Christian faith, as expressed through fasting, her asylum case was ultimately dismissed by the Home Office. In her rejection letter, among other things, the Home Office noted that:

> your previous history shows you can in general live as a Christian born without problems in Pakistan ... as you don't seem to have any religious profile ... Your alleged fear on return is based on threats of persecution from non-state agents and you have not demonstrated they will be able to have any influence over the state ... You claim to have reported both incidents to police stations ... but that these complaints were not fully

investigated. However the evidence you have provided does not demonstrate that you have made efforts to pursue these complaints or take any action regarding the police's failure to investigate … Given that you are a Catholic Christian … it is noted … that there is a strong Catholic community in [city] … it is considered that the size and diversity [of the population] will allow you to relocate with anonymity, it is reasonable to conclude that you will be able to continue practising your religion freely and quietly.

Taken together, the above excerpts show how immigration officials and asylum claimants subscribe to radically different narratives about the nature, extent and even the existence of persecution in Pakistan. One issue is the Home Office's seeming misconceptions on the safety of Pakistanis who are Christian born in a socio-political environment where Christians in general are routinely targeted and abused solely on account of their faith. As one senior legal advisor with experience of dealing with Christians from the Muslim-majority countries commented:

> In terms of the Home Office's point that it's been going okay for so long for Christian born in Pakistan, this is not true for most of the Christians. It's the same thing as swimming in a dangerous stretch of sea every day. You can do it safe for three months, and on three months and one day you drown. … So the fact that someone has been able to practise their Christian faith in an anti-Christian society doesn't mean they are immune to persecution.

We add that what makes EBP useful also makes it limited: it can become detached from the social and political contexts in which persecution occurs, as in this case.

In some cases the sorts of evidence that claimants presented were considered to be low in the hierarchy of evidence. This was mainly the case with documentary evidence such as the first information report (FIR). These are police reports of crimes against the person now claiming asylum. While many of our participants tended to rely on such documents as evidence for their persecution, the Home Office invariably dismissed them as fake. On the whole, on reviewing the rejection letters we found a pattern whereby the Home Office would increasingly refer to expert evidence to paint a broad picture of Pakistanis as fraudulent and opportunistic cheats, and hence potentially bogus claimants. In rejection letters, the Home Office routinely stated that:

> during a presentation at the Ninth European Country of Origin Information Seminar held in Dublin, Ireland, on 26 and 27 May 2004, an Islamabad-based representative of the ... UNHCR ... stated that there is a high level of corruption in Pakistan and that it is possible to obtain many types of fraudulent documents or documents that are fraudulently authenticated by a bona fide stamp or authority.

As Boswell (2009) argues, decision makers often use expert evidence to make their judgements appear neutral as well as to make a justifiable claim to transparency and public acceptance. Our research shows it is possible that an institutional emphasis is emerging in which Pakistanis are regarded as frauds. Such an emphasis may lead to mutual suspicion and prejudice. In this way, asylum adjudication may boil down to assessing the credibility of Pakistanis as a group rather than focusing on individual cases.

The Pakistani Christians we worked with were wary of what they saw as the application of double standards by the Home Office. As one Pakistani pastor put it:

> What puzzles me is, in one context, the Home Office claims that these state agencies are genuine when people say the police did not help me because they are corrupt, but when it comes to evidence, they say the authorities in Pakistan are corrupt. I see this as having double standards.

Perceived bias in relation to immigration interlocutors' personal religious identities

Here we emphasise that the Home Office needs to expand its notion of what counts as evidence, and suggest that this can be achieved by drawing on the knowledge and experience of cultural and religious difference, particularly when dealing with cases involving religiously motivated persecution. As our findings suggest, given the Home Office's limited inclusivity and openness, asylum claimants often lack the confidence and belief that their experiences are taken seriously and listened to by immigration authorities.

For example, the Pakistani Christian asylum seekers we interviewed noted that they often encountered immigration interlocutors from a

Pakistani Muslim heritage. This reflects the UK's diverse religious groups, but some speculate that the Home Office could be deliberately allocating cases involving Pakistani Christians to caseworkers of a Pakistani heritage, presumably for linguistic reasons and assumptions of shared cultural understandings. However, owing to their negative experiences in their country of origin, the Christians we interviewed reported that they often lost the confidence and courage to give evidence of their persecution verbally and defend their asylum cases when faced with individuals whom they perceived to be from the perpetrator group, Pakistani Muslims. Participants routinely drew our attention to their refusal letters, which in most cases included Muslim names as signatories. Often they linked negative asylum decisions to the religious identities of the immigration officials who handled their cases. Such concerns were raised in the context of the Home Office's refusal to accept claimants' requests for non-Muslim caseworkers, on the basis that the system does not keep a database of its employees' religious beliefs.

Claims of religious prejudice and bias were also made about other interlocutors such as interpreters, whom the Home Office regards as mere conduits through which immigration officials and asylum seekers achieve meaningful discourse (see Gib and Good, 2014, for a detailed discussion). The participants alleged that interpreters of similar national heritage but from the Muslim majority were ignorant of appropriate language to describe Christian experiences, and even undermined or manipulated accounts in a discriminatory fashion.

The need for religious diversity, especially in relation to Home Office interlocutors, was succinctly articulated by the Pakistani pastor we cited earlier, as follows:

> In the same way the Home Office is using Muslim Urdu speakers, they should also consider using Urdu speakers who are Christians … or they should at least invite a Christian country expert such as a Pakistani pastor to come and sit there … because the Christian language is not familiar to Muslims. Here in the UK, religion is not given any value. The difference is that in Pakistan, that is an Islamic country, life is about religion … so refugees are so particular as they believe that these Home Office interlocutors are Muslims first, then UKBA workers second.

We suggest that opening up the asylum system through the involvement of, for example, interpreters and experts with knowledge and understanding of the Christian faith and the manner in which it is practised and experienced in Pakistan would help to generate confidence and trust among Christian asylum seekers, perhaps even in cases where a claim is unsuccessful.

Conclusion

Concerns about diversity, community cohesion and the related public fear in the UK of infiltration from Muslim extremists shape both contemporary political discourse and the current restrictive border-control mechanisms. As a result, the Pakistani Christians seeking asylum in the UK may be caught between a rock and a hard place. Initially, the reception experience of Pakistani Christians challenges the neutrality of immigration controls that, in practice, appear to be designed to target immigrants of particular ethnic backgrounds for increased scrutiny. In this context, Pakistani Christians can be subject to misdirected Islamophobia within the immigration and asylum system, given the assumption that the Islamic religion is a core identity of all Pakistani immigrants. Indeed, as Jubany (2011) argues, the tendency to lump together individuals from a particular country/region into a single group seems to be a subculture of the British asylum system as informed by the meta-message of disbelief. This points to the limits to openness and transparency within the UK's supposedly evidence-based immigration policy.

At the same time, in the adjudication of faith-based claims, Pakistani Christians often found themselves confronted with complex obstacles in their endeavour to provide successful evidence of their asylum claims. We have thus sought to make visible the monsters that could be hiding in the UK's evidence-based approaches to determining refugee status, which point to the limited plurality and diversity in the sources and forms of evidence on which the Home Office draws.

Our research has analysed the challenges faced by Pakistani Christians in establishing their Christian identity, and their associated experience of persecution. Our data show that in the absence of good evidence, immigration officials often treat the Christian faith as a mere religious observance (judged from a Western perspective) as opposed to being a core component of one's identity (hence the need for faith to be assessed in the context in which it is practised)

(Nettleship, 2015), requiring officials to be better informed and better trained.

We have also problematised the external evidence or country-of-origin information published by the Home Office, which forms the basis of decisions on Pakistani Christians' claims to asylum. This evidence appears to underestimate the extent and forms of persecution experienced by Christians in Pakistan. For these Christians, both in Pakistan and in many other Muslim-majority countries, faith not only informs their identity but shapes all aspects of their private and public life. This reflects both the way in which Christians themselves experience their faith and also how they are regarded and treated by non-Christians (as employers, neighbours, the police and the judiciary).

Finally, we have drawn attention to what appears to be a lack of religious diversity in the immigration interlocutors (though in a supposedly religiously neutral asylum system) and its impact upon the ability of Pakistani Christians to defend their claims verbally. Our participants routinely made claims of religious discrimination and bias in a context where their asylum cases are frequently handled and facilitated by caseworkers and interpreters of Pakistani Muslim heritage. We see this chapter as filling a significant gap, not only in terms of evidence, but also in current research and public debates on asylum seekers from Muslim-majority countries.

References

Ahmed, S. (2007). A phenomenology of whiteness. *Feminist Theory*, 8(2), 149–168.

Anderson, B. (2013). *Us and Them? The Dangerous Politics of Immigration Control*. Oxford: Oxford University Press.

Boswell, C. (2009). *The Political Uses of Expert Knowledge: Immigration Policy and Social Research*. Cambridge: Cambridge University Press.

Botterill, L., and Hindmoor, A. (2012). Turtles all the way down: Bounded rationality in an evidence-based age. *Policy Studies*, 33(5), 367–379.

Cameron, D. (2011). PM's speech at the Munich Security Conference. *Gov. uk*, Cabinet Office, Prime Minister's Office, 10 Downing Street. Retrieved 20 June 2013 from: www.number10.gov.uk/news/pms-speech-at-munich-security-conference.

Clarence, E. (2002). Technocracy reinvented: The new evidence based policy movement. *Public Policy and Administration*, 17(3), 1–11.

Dinham, A., Furbey, R., and Lowndes, V. (eds) (2009). *Faith in the Public Realm: Controversies, Policies and Practices*. Bristol: Policy Press.

Gibb, R., and Good, A. (2014). Interpretation, translation and intercultural communication in refugee status determination procedures in the UK and France. *Language and Intercultural Communication*, 14(3), 385–399.

Green, D. (2010). Immigration: Damian Green's speech to the Royal Commonwealth Society. *Gov.uk*, UK Visas and Immigration, Immigration Enforcement. Retrieved 15 April 2013 from: www.gov.uk/government/speeches/immigration-damian-greens-speech-to-the-royal-commonwealth-society.

Home Affairs Committee (2013). *Asylum: Seventh Report of Sessions 2013–2014*. House of Commons. London: The Stationery Office Limited.

Home Office (2015a). National statistics: Asylum. *Gov.uk*, Home Office, Immigration statistics. Retrieved 14 December 2015 from: www.gov.uk/government/publications/immigration-statistics-april-to-june-2015/asylum.

Home Office (2015b). Asylum policy instruction: Assessing credibility and refugee status. *Gov.uk*, UK Visas and Immigration. Retrieved 12 February 2015 from: www.gov.uk/government/publications/considering-asylum-claims-and-assessing-credibility-instruction.

Home Office (2015c). *Country Information and Guidance. Pakistan: Christians and Christian Converts*. Home Office. Retrieved 14 March 2015 from: www.gov.uk/government/uploads/system/uploads/attachment_data/file/402591/cig_pakistan_christians_and_christian_converts_v1_0_2015_02_10.pdf.

Joppke, C., and Torpey, J. (2013). *Legal Integration of Islam: A Transatlantic Comparison*. Cambridge, MA: Harvard University Press.

Jubany, O. (2011). Constructing truths in a culture of disbelief: Understanding asylum screening from within. *International Sociology*, 26(1), 74–94.

Kagan, M. (2010). Refugee credibility assessment and the 'religious imposter' problem: A case study of Eritrean Pentecostal claims in Egypt. *Vanderbilt Journal of Transnational Law*, 43(5), 1179–1232.

Kisby, B. (2011). Interpretations, facts, verifying interpretations: Public policy, truth and evidence. *Public Policy Administration*, 26(1), 107–127.

Lentin, A. (2011). What happens to anti-racism when we are post-race? *Feminist Legal Studies*, 19(2), 159.

Lowndes, V. (2016). Narrative and story-telling. In G. Stoker and M. Evans (eds), *Evidence-Based Policymaking in the Social Sciences: Methods that Matter*. Bristol: Policy Press.

Majone, G. (1989). *Evidence, Argument and Persuasion in the Policy Process*. New Haven, CT: Yale University Press.

Nettleship, P. (2015). All-Party Parliamentary Group for International Freedom of Religion or Belief: Pakistani evidence hearings submission. *APPG*, Pakistan

report. Retrieved 20 February 2016 from: https://freedomdeclared.org/in-parliament/pakistan-report/.
Pearce, W., and Raman, S. (2014). The new randomised controlled trials (RCT) movement in public policy: Challenges of epistemic governance. *Policy Sciences*, 47(4), 387–402.
Souter, S. (2011). A culture of disbelief or denial? Critiquing refugee status determination in the United Kingdom. *Oxford Monitor of Forced Migration*, 1(1), 48–59.
Spencer, S. (2011). *The Immigration Debate*. Bristol: University of Bristol, Policy Press.
Statham, P., and Tillie, J. (2016). Muslims in their European societies of settlement: A comparative agenda for empirical research on socio-cultural integration across countries and groups. *Journal of Ethnic and Migration Studies*, 42(2), 177–196.
Thomas, R. (2006). Assessing the credibility of asylum claims: EU and UK approaches examined. *European Journal of Migration and Law*, 8, 79–96.
UNHCR (1998). *Note on Burden and Standard of Proof in Refugee Claims*. Geneva: UNHCR.
UNHCR (2004). *Guidelines on International Protection: Religion-Based Refugee Claims under Article 1A(2) of the 1951 Convention and/or the 1967 Protocol Relating to the Status of Refugees*. Geneva: UNHCR.
UNHCR (2013). *Beyond Proof: Credibility Assessment in EU Asylum Systems*. Brussels: UNHCR.
Wesselink, A., Colebatch, H., and Pearce, W. (2014). Evidence and policy: Discourses, meanings and practices. *Policy Sciences*, 47, 339–344.
Yin, R. (2003). *Applications of Case Study Research*. Thousand Oaks, CA: Sage Publications.

Part II
Responsibility

5

Responsibility

Barbara Prainsack, Sabina Leonelli

Openness has become fashionable. Governments, software and even humans are furnished with the adjective 'open'. This is not a quiet and modest adjective, but entails a demanding cluster of requirements. To be open means to be transparent, responsible, accountable and inclusive. In other words: to be open is to be good.[1]

What does this mean for science? If we understand openness as the commitment to make the tools and processes of science replicable and open to scrutiny, then science has had a particularly close relationship with openness from its earliest beginnings. It is true that science has been, and still is, to a large extent an elite activity. But the idea that people other than those involved in the creation of a scientific finding need to be able to scrutinise claims in order to find possible errors and to corroborate or falsify hypotheses and claims is intrinsic to the very concept of science. It is the requirement of transparency, intelligibility (if only to peers) and openness to scrutiny that distinguishes science from other instances of skilful practice or from an informed debate. Openness about how we arrive at conclusions on the basis of evidence is what enables the type of empirical self-corrective knowledge creation that science has (often rightly) claimed to be.

Why is it, then, that openness – in science, but also in other domains of life – has become such a buzzword in the twenty-first century? There

---

[1] Dave Eggers's 2013 novel *The Circle* is a stark illustration of this, particularly since it was meant as science fiction and yet seems to describe the cult of transparency as a solution to social problems.

are several answers to this question. One answer lies in our political economy. The shift from familial to corporate capitalism (Fraser, 2015) and the financialisation of capitalism have together solidified the dominance of commercial interests over politics. Corporate actors have easy access to national power centres, to the extent that they co-regulate important national policies (Gamble, 2014). As a result, not only governments but also citizens have lost control over important policy domains such as housing, work and energy (Wagenaar, 2016). Social inequalities have been reified; even most left-wing progressive movements no longer see them as something that needs to be abolished but instead as something to be managed. Within this political economy, openness – particularly when interpreted as an appeal to transparency as means to unmask inequalities and corruption – assumes the role that gallantry had in the eighteenth and nineteenth centuries: while seemingly easing the relationship between women and men, but not also between the rich and the poor, it can also be associated with a system that reinforces underlying inequalities and benefits one group at the expense of the other.

Making science open bears the connotation of democracy and equality, a bit as has been the case with the internet. There has been an assumption for some time that, because everybody in principle has equal access to web-based tools, the internet will make societies more democratic. As is well known by now, the supposed openness and accessibility of the internet has indeed given voice to some people who would not have been heard otherwise, but it has also given powerful actors even more power (Taylor, 2014). It has become clear in many contexts that fostering transparency does not always foster meaning making and intelligibility. Sometimes it may instead draw attention away from all the obstacles to equality and democracy that stem from the dominant global and socio-political order.

Another answer is rooted in the development and proliferation of digital tools and instruments to produce data. Once data are available in digital form, three key questions arise, either explicitly or tacitly: who owns them, who should have access to them and who should be allowed to use them. This is true for any digital data set – including our postings on social media, billing data held by health insurance companies or geolocation information collected by mobile phone companies. Scientific data, however, are under particular pressure to be available, intelligible and usable to a wide range of users, because of the expectation that knowledge production (particularly that sponsored by public money) should benefit society at large and be accessible by citizens at all times and in all available formats. There is an emerging consensus that, wherever ethically and legally possible, research data should not only be open to any interested scientist but also to the general public as the funders and stakeholders of science. At the same time, the expertise required to make data open in the sense of intelligible and reusable is often disregarded, with several open-data initiatives targeting the general public but providing few tools for them to meaningfully engage with the data and extract information from them.

There are other answers to the question why the notion of openness has assumed so much cultural importance, which we have discussed elsewhere (e.g. Leonelli, 2016; Levin and Leonelli, 2016; Prainsack, forthcoming). What the examples that we have given here show is that open science is a political project to an even greater extent than it is a technological one. The chapters in this part illustrate the range of these political aspects and how they may intersect and converge with scientific, technological and organisational concerns in the running of large research initiatives. In the remainder of this introduction we will discuss some of the themes and tensions that the chapters highlight and bring them in dialogue with the political and epistemic characteristics of the open-science movement.

Centralisation and epistemic diversity

Whenever they try to disseminate information and resources, including data, models, specimens, software and papers, beyond their immediate work circles, researchers encounter issues of standardisation (Bowker and Star, 1999). There is a strong and often unresolvable tension between attempting to preserve the diversity of practices that constitutes the epistemic richness of research and its outputs, and setting up common venues and formats where outputs can be shared, discussed and reused (Leonelli, 2016). Such difficulties become particularly jarring when attempting to manage and distribute information, and foster participation in massive multiheaded institutions and projects, where a strongly centralised structure is combined with multiple approaches and types of practice.

Chapter 6 by Eleanor Hadley Kershaw highlights this beautifully by discussing Future Earth, one of the most complex bodies of individuals, resources, tools, approaches, groups, institutions and data ever assembled to confront a broad and all-encompassing set of environmental and social challenges across the globe as a whole. The ambitious scope and remit set for Future Earth by its governing council (comprising the International Council for Science (ICSU), the Belmont Forum and other international partners) is reminiscent of the open-science goals of bringing together all potential stakeholders and sharing knowledge and tools widely for the greater good. This is justified, since addressing global challenges arguably requires as inclusive and pluralistic an approach as possible, and thus seeks to combine efforts by experts in all fields and methods, capitalising on the variety of existing skills and knowledge. At the same time, this huge scope has resulted in an organisation that has little coherence or cohesion. Attempts to unify or even just coordinate approaches can and often do fail, as the multiplicity of challenges and approaches involved (which go well beyond disciplines, since interdisciplinary perspectives and innovative methods are typically very useful to such an endeavour) defy easy standardisation and harmonisation across participants. The inability to resolve the tensions outlined by Hadley Kershaw, particularly the one between the will to harmonise and cooperate and the importance of respecting and preserving diversity, risk making Future Earth a monstrous enterprise that nobody understands in its entirety, and in which operations are disjointed and out of synch with one another. At the same time, as argued by Hadley Kershaw, such tensions and differences in interpretation could be seen as productive openings for negotiation.

In sum, this case in turn raises questions about the relation between openness and pluralism in approaches (and, particularly, disciplines), given the various ways in which researchers working in different fields and locations understand the notion of openness, the challenges and opportunities that it can bring, and the ways in which it can be instantiated (Levin et al., 2016; Tenopir et al., 2015).

Inclusiveness and social justice

Another type of pluralism, which is a motor for openness but also a potential obstacle to sharing and exchanging data, is geographical and cultural. Chapter 7 by Alison Mohr on energy transitions research brings the power differentials and diversity of location and research environment between the North and the South into sharp relief, thus highlighting the challenges of social justice and spatial positioning that underlie attempts to apply the notion of openness across national boundaries. Mohr emphasises how translating Northern ideas into global South contexts involves deep structural changes to the latter, which are difficult to implement and whose systemic implications are hard to predict and manage. And, indeed, the situation that she describes parallels attempts to construct digital infrastructures for open-science projects that benefit the developing world. Whether such infrastructures target citizen scientists or professional researchers, they are often devised and developed by individuals based in high-income countries, which makes it difficult for prospective users in low-income countries to share the same assumptions and conditions for entry. Open science can certainly serve to enhance social justice by fostering better exchange of data, better quality control of the information data exchanged, and cooperation in the ways in which research is organised (e.g. by helping researchers interested in exploring the same phenomenon to find each other and network). Open-science initiatives can also enhance the social, cultural and economic effects of research outputs on the various publics who, directly or indirectly, engage with them or are affected by them (e.g. how participants think about the impact of their work, what are their publics, what and whom they address). However, to reach these goals, openness advocates need to take account of the staggering differences between research environments in various parts of the world, which include not only available funding and resources, but also the wider infrastructure and institutionalisation of scientific work (Bezuidenhout et al., 2016). Many open-science projects, no matter how well intentioned, do not treat their prospective participants as equals and tend to impose their own vision of what it means to do research and share information without taking account (and in many cases, without being fully aware of) the conditions under which some of their publics and users work. As Chris Kelty (2008) has argued in his study of the development of the open-source movement, an effective way to engage in open research is to create one's own public through deep engagement and the distribution of stakes and responsibilities, so as to reflexively identify and take into account existing inequalities – an approach that some citizen-science initiatives are already successfully using.

Transparency

Both Hadley Kershaw and Mohr address the scale of nested international structures, disciplines and institutions at global and local levels. Within such systems it is not clear that making knowledge open, in the sense of making everything accessible (for instance via online dissemination), enhances the transparency of the efforts at hand in a meaningful way. In many cases, understanding a given intervention or claim requires expertise and familiarity with the issues. Furthermore, the sheer quantity and scale of information that can be made available can baffle and confuse readers, rather than providing them with useful information (Floridi, 2014). Too often is the offer of unlimited information accompanied by interpretative moves left unacknowledged, such as with WikiLeaks (Rappert, 2015). In many instances people who make data accessible to others are not even aware of the assumptions and interpretations that have become part of the data sets, if only by ordering and classifying them (Bowker and Star, 1999). In sum, obtaining transparency is not merely a matter of making information available but rather a matter of explicating one's own assumptions, providing critical tools to unravel complex claims or situations, and engaging with them critically.

Responsible Innovation

In the 1960s and 1970s several organisations and initiatives in Europe and in North America sought to 'radicalise' science (Bell, 2013; Rose and Rose, 1976, 1979). Fuelled by the experience of how science had put itself in the service of imperialistic and martial interests, culminating in the atrocities of World War II, initiatives associated with the radical-science movement challenged the perception that science is something neutral, something independent of economic and political interests. The radical-science movement was not an attempt by people to tame or even govern science from the outside. On the contrary, it was driven by professional scientists. Organisations such as Scientists and Engineers for Political and Social Action in the USA and the British Society for Social Responsibility in Science in the UK argued that science should be seen and analysed as an integral part of a particular social and political order. Responsible science thus meant that professional scientists and publics alike reflected critically on the questions that science asked, the methods it used and the resources it obtained.

For many national and international organisations today, responsibility in science has a very different meaning. Politics has largely been sucked out of it. Responsibility has mostly become synonymous with complying with ethical codes. Also, most of the 'radical' scientists of today, including citizen scientists and those who are self-proclaimed proponents of 'forbidden' research (e.g. MIT Media Lab, 2016), are no longer asking the big question of what kind of society we want (see also Delfanti, 2013). While some of them – among them, for example, do-it-yourself biologists – do challenge the norms and goals of institutionalised science, which they see as compromised by commercial interests, many of them help to make science 'better' within the epistemic commitments and rationale of institutionalised professional science (see Irwin, 2015; Prainsack and Riesch, 2016). They often fill the void that public actors have left, for example, by crowd-funding scientific research or by undertaking tasks that would be too onerous or too difficult for professional scientists to do (Del Savio et al., 2016; Prainsack, 2014). Here, responsibility no longer means a collective responsibility to ensure that science contributes to making our societies more just and more dignified for everybody to live in. Instead, it increasingly connotes a duty for individual scientists – professional scientists or citizen scientists – to be useful to existing systems.

In its worst instantiation – and unfortunately this is the case for the EU's pet concept of responsible research and innovation (RRI) – responsibility boils down to the duty to create a societal impact, which in turn is seen as an incentive – or even a societal need – to partner with industry. In its less depressing instantiations, contemporary interpretations of responsibility in science include the duty to think carefully about its societal impacts and to increase inclusiveness (e.g. Kreissl et al., 2015). Viewed in this context, Stevienna de Saille and Paul Martin's approach in chapter 8 to analyse RRI using the notion of the monster is both intriguing and useful. Drawing upon monster theory (e.g. Smits, 2006) these authors argue that an entity's monstrosity is not so much determined by the extent to which it is seen as infringing supposedly natural principles, as it is shaped by its inhabiting mutually exclusive categories (see pp. 152–153). To de Saille and Martin, RRI represents both the fear of monsters and the fear of what they call the 'monstrous regiment' (see p. 150). According to the authors, this is because RRI can be seen as serving two mutually exclusive goals. The first is to generate private wealth while socialising public risk, and the second is to involve publics and give them the chance to 'destroy the new market' (see p. 156). While this is an original way of analysing RRI, it may not be the only way in which RRI can be seen in the current political and economic landscape. Following scholars such as Slavoj Žižek (2006) and Philip Mirowski (2014), what sets the current situation apart from previous eras is that the goals of economic growth, wealth creation and creating public benefits are seeing as entirely compatible.[2] This has become possible because our societies have naturalised social inequalities and injustice. RRI could be seen as one of the tools that help to manage such inequalities.

Conclusion

Openness is not more important today than it has been in earlier decades. But the wide variety of goals and uses to which this notion is now routinely subjected has generated a complex nexus of expectations and associations. Many of these expectations fail to materialise in reality, or they do so only for a highly selected public and in ways that only partly reflect the intended aims. That the various meanings and functions of openness are often flattened in scientific and public discourse is unhelpful here, as it suggests that questions about the success or value of openness can be answered with a simple 'yes' or 'no'.

Some of the tensions and challenges pertaining to open science that are addressed in this part – and in this volume more broadly – cannot be resolved very easily; in fact, some of them may not be soluble at all, because they reflect the fact that openness can be interpreted in many different ways, and yet each interpretation can work to the benefit or at the expense of others. The lesson to draw from this is that, rather than treating openness as a term that describes the value system or political commitments of an institution or practice, we must take a

---

2 The current boom of concepts such as effective altruism (effectivealtruism.org) and practices such as social-impact investing (e.g. Bugg-Levine and Emerson, 2011) are but two examples of this.

deeper look at the values, goals and purposes that openness supports and realises in concrete instances. The contributions by Mohr, Hadley Kershaw, and de Saille and Martin all do this in different ways and with different results, once again illustrating the breadth and partly contradictory nature of practices and institutions of openness.

References

Bell, A. (2013). Beneath the white coat: The radical science movement. *Guardian*, 18 July. Retrieved 11 August 2016 from: www.theguardian.com/science/political-science/2013/jul/18/beneath-white-coat-radical-science-movement.

Bezuidenhout, L., Rappert, B., Kelly, A., and Leonelli, S. (2016). Beyond the digital divide: Towards a situated approach to open data. *Science and Public Policy*, 13 July 2017, scw036 [epub ahead of print], doi: 10.1093/scipol/scw036.

Bowker, G. C., and Star, S. L. (1999). *Sorting Things Out*. Cambridge, MA: MIT Press.

Bugg-Levine, A. and Emerson, J. (2011). Impact investing: transforming how we make money while making a difference. *Innovations*, 6(3), 9–18.

Delfanti, A. (2013). *Biohackers: The Politics of Open Science*. London: Pluto Press.

Del Savio, L., Prainsack, B., and Buyx, A. (2016). Crowdsourcing the human gut. *Journal of Science Communication*, 15(3), A3.

Floridi, L. (2014). *The Fourth Revolution: How the Infosphere is Reshaping Human Reality*. Oxford: Oxford University Press.

Fraser, S. (2015). *The Age of Acquiescence: The Life and Death of American Resistance to Organized Wealth and Power*. London: Hachette.

Gamble, A. (2014). *Crisis Without End? The Unravelling of Western Prosperity*. Basingstoke: Palgrave Macmillan.

Irwin, A. (2015). Public engagement with science. In J. D. Wright (ed.), *International Encyclopedia of Social and Behavioral Sciences*, 2nd edition (pp. 255–260). Oxford: Elsevier.

Kelty, C. (2008). *Two Bits*. Durham, NC: Duke University Press.

Kreissl, R., Fritz, F., and Ostermeier, L. (2015). Societal impact assessment. In J. D. Wright (ed.), *International Encyclopedia of the Social and Behavioral Sciences*, 2nd edition (pp. 873–877). Oxford: Elsevier.

Leonelli, S. (2016). *Data-centric Biology: A Philosophical Study*. Chicago, IL: Chicago University Press.

Levin, N., and Leonelli, S. (2016). How does one 'open' science? Questions of value in biological research. *Science, Technology and Human Values*, 42(2), 280–305 [epub ahead of print], doi: 10.1177/0162243916672071.

Levin, N., Leonelli, S., Weckowska, D., Castle, D., and Dupré, J. (2016). How do scientists define openness? Exploring the relationship between open science policies and research practice. *Bulletin of Science and Technology Studies*, 36(2), 128–141.

Mirowski, P. (2014). *Never Let a Serious Crisis Go to Waste: How Neoliberalism Survived the Financial Meltdown*. London: Verso.

MIT Media Lab (2016). Forbidden research. *MIT News*, 25 July. Retrieved 12 August 2016 from: http://news.mit.edu/2016/forbidden-research-media-lab-0725.

Prainsack, B. (2014). Understanding participation: The 'citizen science' of genetics. In B. Prainsack, S. Schicktanz and G. Werner-Felmayer (eds), *Genetics as Social Practice: Transdisciplinary Views on Science and Culture*. Farnham: Ashgate.

Prainsack, B. (forthcoming). *Personalization from Below: Participatory Medicine in the 21st Century?* New York: New York University Press.

Prainsack, B., and Riesch, H. (2016). Interdisciplinarity reloaded? Drawing lessons from 'citizen science'. In S. F. Frickel, M. Albert and B. Prainsack (eds), *Investigating Interdisciplinary Collaboration: Theory and Practice Across Disciplines*. New Brunswick, NJ: Rutgers University Press.

Rappert, B. (2015). Sensing absence. In B. Rappert and B. Balmer (eds), *Absence in Science, Security and Policy* (pp. 3–33). London: Palgrave.

Rose, H., and Rose, S. (1976). The radicalisation of science. In H. Rose and S. Rose (eds), *The Radicalisation of Science: Ideology of/in the Natural Sciences* (pp. 1–31). London: Macmillan Education.

Rose, H., and Rose, S. (1979). Radical science and its enemies. In R. Miliband and J. Saville (eds), *Socialist Register* (pp. 317–335). Atlantic Highlands, NJ: Humanities Press.

Smits, M. (2006). Taming monsters: The cultural domestication of new technology. *Technology in Society*, 28(4), 489–504.

Taylor, A. (2014). *The People's Platform: Taking Back Power and Culture in the Digital Age*. New York: Picador.

Tenopir, C., Dalton, E. D., Allard, S., Frame, M., Pjesivac, I., Birch, B., Pollock, D., et al. (2015). Changes in data sharing and data reuse practices and perceptions among scientists worldwide. *PLoS One*. Retrieved 27 November 2016 from: http://dx.doi.org/10.1371/journal.pone.0134826.

Wagenaar, H. (2016). Democratic transfer: Everyday neoliberalism, hegemony and the prospects for democratic renewal. In J. Edelenbos and I. van Meerkerk (eds), *Critical Reflections on Interactive Governance* (pp. 93–119). Cheltenham: Edward Elgar.

Žižek, S. (2006). Nobody has to be vile. *London Review of Books*, 28(7), 10.

6

Leviathan and the hybrid network: Future Earth, co-production and the experimental life of a global institution

Eleanor Hadley Kershaw

In the opening words of *A Sociology of Monsters*, John Law caricatures a middle-class white male, middle-aged, non-disabled person's perspective on the history of sociology: 'We founded ourselves on class; then, at a much later date we learned a little about ethnicity; more recently we discovered gender; and more recently still we learned something ... about age and disability' (Law, 1991: 1). Thus, the hypothetical sociologist's gradual realisation that ' "his" sociology had never spoken for "us": that all along the sociological "we" was a Leviathan that had achieved its (sense of) order by usurping or silencing the other voices' (Law, 1991: 1). In acknowledging the struggles – the 'pain' – of sociology's gradually expanding and more inclusive scope, Law signals the (potential) monstrosity of discipline and exclusivity in the formation and perpetuation of academic knowledge and communities.

Whether or not one subscribes to this view of sociology as monstrous, science and technology studies (STS) is replete with monsters. From Haraway's cyborg (1991) to Latour's appeal for us to love and care for our technologies rather than abandon them as Frankenstein did his creation (Latour, 2011), the central preoccupations of this field concern the (blurred) categories of and relations between nature; culture; the human, non-human or more-than-human; the scientific; technological; social; material; epistemic and the political: the construction, maintenance and disturbance of our 'natural' and 'social' orders and kinds. Whether leviathan in the sense of biblical sea monster, monolithic and powerful organisation, or Hobbes's body politic ruled by the sovereign head of state – or whether ambiguous hybrid – the monster is a useful metaphor and heuristic for exploring and describing disruptions to order and control; the unusual, unfamiliar or potentially threatening; and our anticipation, fears and hopes.

With this in mind, might disciplines, research communities and knowledge domains be usefully characterised as leviathans, giant structures composed of many individuals governed by one singular centralised agency policing their boundaries? Or are they best seen as hybrid networks, configurations or assemblages of the scientific and the political, the material and the epistemic, the social and the natural? Or perhaps it is more productive to consider our epistemic apparatus not as a monster itself, but as facing monsters in the process of its continuous (re)formulation, (re)definition and (re)structuring.

This chapter explores efforts to bring about transformations in global environmental change (GEC) research institutions, communities and cultures. In particular, it focuses on the reconfiguration of several existing international GEC research programmes into one initiative: Future Earth (henceforth FE), an international research initiative on GEC and sustainability that was launched in 2012 and became fully operational in 2015. This reorganisation is accompanied (and in part motivated) by ambitions for a 'new type of science' (FE, 2014: 2) and 'a new "social contract" between science and society' (FE, 2013: 11). To achieve these aims, FE is unique in explicitly adopting co-design and co-production of knowledge as a strategic agenda for the governance, coordination and production of research from global to local levels.

Here, co-design and co-production (and transdisciplinarity) can be understood as the opening up of science (or research more broadly) and its governance to the participation of a wider range of stakeholders, with the goal of transforming the ways it is done and used, as well as its efficacy and impact. However, we do not yet know how this transformation might be enacted in practice in a concrete programme. Given this, what are constructive ways of understanding this initiative and others like it? Are there monsters lurking in the uncharted territories of the deliberate co-production of new research at a global scale? And what is FE – a powerful leviathanesque organisation, a hybrid network, a combination of the two or something else entirely? This chapter, written in early 2016, explores the answers to these questions at a particular moment in FE's development, drawing on data from a qualitative case study of FE conducted between 2013 and 2015. FE has since evolved (and will continue to evolve for its duration), but may still face many of the questions and tensions considered here.

The following section provides background on transformations in research systems and cultures, changing science–society relations, and a brief introduction to the concepts of co-design, co-production and transdisciplinarity.

Changing research systems, cultures and science–society relations

Since the 1980s and 1990s, scholars of research policy and STS have explored changes in research systems, institutions, philosophies, cultures and practices. The various conceptualisations of transformations in the dynamics between research and society include post-normal science (Funtowicz and Ravetz, 1993), mode 2 knowledge production (Gibbons et al., 1994) and post-academic science (Ziman, 2000), amongst many others. While such concepts are diverse, they all describe, theorise or advocate a shift from the disciplinary organisation of research within the academy towards a greater recognition of socio-economic priorities and involvement of broader communities – particularly non-academic actors – in the governance and production of research, whether at a local, national, regional or global scale (Hessels and Van Lente, 2008). These ideas and accounts are not uncontested, but some of them have also taken on a performative role, both suggesting that a new organisation of knowledge production should take place (or is taking place) and simultaneously participating in its realisation (Godin, 1998).

Alongside the emergence of these ideas there have been increasing calls within national and international research (funding) communities for multi-, inter- and transdisciplinarity; public and stakeholder engagement; and RRI – often to address grand societal challenges such as global environmental change, health or sustainability (Felt and Wynne, 2007). In this context, co-design and co-production have been advocated as research and/or governance approaches – for example, by UK research councils and academics (Campbell and Vanderhoven, 2016), as well as in wider public policy contexts (Durose and Richardson, 2015).

However, while these initiatives aim to move beyond the traditional disciplinary organisation of research and open up research and policy processes to a wider range of actors, the concepts underpinning them retain a high degree of interpretive flexibility (Ribeiro et al., 2016), sometimes serving as 'buzzwords' that mobilise people and resources (Bensaud Vincent, 2014), or as 'boundary objects' that enable diverse parties with different perspectives to work together (Star and Griesemer, 1989). In particular, co-production and related concepts such as co-design and transdisciplinarity have various meanings in different theory and practice contexts, including STS, public policy, environmental and sustainability research, health, community and development studies, and the arts and humanities. In general, these terms refer to the participation or collaboration of non-academic actors in research (whether defining questions, gathering or analysing data, or engagement in other aspects of the research process) (e.g. Pohl et al., 2010), or the involvement of communities and/or other non-governmental actors in policy processes or public service provision (Bovaird, 2007). Co-production is also employed as an analytical idiom by Sheila Jasanoff and other scholars in STS to signify the co-constitution of science (knowledge) and social (political) order (Jasanoff, 2004).

While the language of STS and cognate fields (e.g. 'participation', 'engagement', 'co-production') is adopted in research governance, practice and broader policy contexts, the motivations for using these approaches in these contexts do not always follow the original logics espoused by scholars, and implementing the concepts can be challenging (Irwin, 2006; Pohl et al., 2010; Stirling, 2008). The reasons for proposing and adopting these approaches to reconfiguring epistemic practices, knowledge domains, and science–society relations are at least as diverse as the range of interpretations of the concepts themselves. STS scholars continue to argue for social agency, for opening up or democratising science and governance, and ensuring accountability of science/scientists and policy makers to broader society. But they also suggest that, despite the proliferation of the language of STS, deficit and linear models persist that assume one-way relationships between experts and other stakeholders, and between knowledge and socially beneficial outcomes via policy, failing to acknowledge the uncertainties and limits of knowledge and governance (Jasanoff, 2003; Stirling, 2008; and see chapter 12).

In the case of the co-production of research, even within STS there are varying rationales for advocating co-production practices, driven by different logics (Lövbrand, 2011, following Barry et al., 2008). One rationale is underpinned by a 'logic of ontology' that hopes to challenge the broad questions of collective purpose and ontological assumptions inherent in particular world views or issue framings. Another is a 'logic of accountability' that calls for the alignment of research portfolios with societal need. In the context of a research project on EU climate policy, these different rationales seemed incompatible. Broadly speaking, policy makers in the co-production exercise wanted knowledge that would be 'useful' within a predefined framework, while social scientists wanted to challenge the policy framework.

Other initiatives to implement or study co-production have tended to focus on the intentional pursuit of co-production at local, national or regional levels and in individual projects and organisations (e.g. Robinson and Tansey, 2006), or the broader (often much less explicit) processes of the co-constitution of science and social order in national and international contexts (e.g. Beck and Forsyth, 2015). However, only a few studies have examined the adoption of co-design and co-production as principles for the governance and production of new research at a global scale (Lahsen, 2016; van der Hel, 2016). This chapter addresses that gap by exploring how co-production is understood – and how it relates to broader questions of institutional identity and function – in the case of FE.

Future Earth background

FE, a major international research initiative on global environmental change (GEC) for sustainability, was officially launched by the Science and Technology Alliance for Global Sustainability in June 2012, during the UN Conference on Sustainable Development (Rio+20).[1] Its unique ambition is to provide a global framework and international coordination of new research on GEC and sustainability (as opposed to the

1 The Alliance (formed in 2010) comprises the International Council for Science (ICSU); the International Social Science Council; the Belmont Forum of global change research funders; the United Nations Educational, Scientific and Cultural Organization (UNESCO); the United Nations Environment Programme; the United Nations University; and the World Meteorological Organization (see www.stalliance.org/).

large-scale synthesis of existing research in assessments undertaken by the Intergovernmental Panel on Climate Change (IPCC) and the Intergovernmental Platform on Biodiversity and Ecosystem Services (IPBES)). It aims to become 'a major international research platform providing the knowledge and support to accelerate our transformations to a sustainable world' and 'a platform for international engagement to ensure that knowledge is generated in partnership with society and users of science', merging several existing international GEC research programmes (FE, n.d. a).

These programmes aimed to coordinate and promote international interdisciplinary research on GEC, set research agendas, and make links between research and policy (FE, 2013: 13). Between 2006 and 2009 the International Council for Science (ICSU) reviewed some of the programmes, concluding that there was a 'need to implement a single strategic framework for Earth system research in the near future' because the research landscape was 'complex, confusing, and often [led] to inefficient use of human, institutional, and financial resources' (ICSU, 2009: 2). In 2009 ICSU and the other co-sponsors began a visioning process to rethink the future of the programmes (which converged with the development of the 'Belmont Challenge', a funders' vision for GEC research (Belmont Forum, 2011)), and later established a transition team to design a new ten-year initiative to succeed them. FE was launched in June 2012, and became fully operational in 2015 through the change from an interim to a permanent secretariat. Key features planned for the fully operational phase of FE included:

- an emphasis on integrated research across disciplines spanning natural and social science, the humanities, and engineering
- the co-design and co-production of research with stakeholder groups including funders, policymakers, civil society and business
- the initiative's global scope, encompassing all regions, and bottom-up input from the research community and other stakeholders
- the accelerated delivery of solutions-oriented, policy-relevant research (FE, 2013; ICSU, n.d.).

The co-design and co-production of relevant knowledge in particular are billed as 'one of the most innovative aspects' of the initiative (FE, n.d. b). FE thus exemplifies calls for transformations in research systems and knowledge-making communities towards engagement with non-academic stakeholders. But what do co-design and co-production mean in FE, and why have they been adopted as core features of the initiative? And how can we make sense of this type of programme?

Jasanoff and Wynne (1998: 58–59) note that two potential avenues for development were mooted for one of FE's precursors, the International Geosphere–Biosphere Programme (IGBP), in the mid-1990s (Barron, 1994). Firstly, the 'rich tapestry' model, in which diverse (sometimes pre-existing) national and individual contributions or initiatives would have been grouped under the broad umbrella of the IGBP. Here, IGBP's role would have been to facilitate communication and collaboration, but its identity, visibility and ability to attract funding may have been weakened. The second model was that of the 'flagship', where effort would have been focused on 'task-oriented activity' (such as major field experiments or modelling activities) with closer integration, greater harmonisation and stricter dirigisme, reinforcing IGBP's status as an autonomous international programme at the cost of its inclusivity and diversity. In terms of the metaphorical monsters described at the start of this chapter, these models respectively align with the notion of a knowledge community as a hybrid science–society network, and as a monstrous leviathan.

FE faces similar dilemmas in relation to its structure and role. Might it become – or is it best viewed as – a loose hybrid network of the scientific–political, the material–epistemic and the social–natural, in which there may always be ambiguity around what type of organisation it is? Or is it better viewed as a leviathan, a giant monolithic structure composed of many individuals governed by one centralised agency aiming to coordinate, prescribe, direct, harmonise and standardise knowledge production, potentially to the detriment or exclusion of some voices? Or is it both – or neither? And what role might co-design and co-production play in these various scenarios?

Emerging tensions and ambiguities

Drawing on a qualitative case study of FE conducted between 2013 and 2015, this section firstly explores the diverse designations and roles imagined for FE, noting that ambiguities between them – and in the ambitious remit of FE – were seen as potentially problematic by some FE and external actors. Six specific tensions and ambiguities arise from these diverse visions and are likely to proliferate as FE is extended and replicated at different scales and in different contexts. Ambiguities in co-design and co-production in FE are then identified, starting with the predominant rationale for this approach (co-production for utility) before detailing two further views (co-production for democratisation, and co-production as a threat). The necessity of resolving these ambiguities is questioned, and their relation to the flagship/leviathan and rich-tapestry/hybrid-network models is discussed.

During the study, FE was still very much an organisation in the making; many aspects of its design and implementation were still ambiguous and yet to be determined. Data collection for this study was concluded before the transition from the interim to the permanent secretariat in 2015. FE has since evolved and will continue to evolve for its duration.

Visions of Future Earth and the negotiation of ambiguities

During the period of study, visions of FE – of what it is supposed to be and what roles and functions it is supposed to have – were diverse, ambitious and, sometimes, ambiguous. In official and internal documents and webpages (produced between 2010 and early 2015), as well as in interviews and focus groups conducted as part of this study,[2] FE has been conceptualised in a range of ways. Some of these visions evoke a leviathanesque 'flagship', for example: 'a consolidated and comprehensive effort, a flagship initiative on Earth System Research for Global Sustainability' (ICSU, 2011); 'the global research platform providing the knowledge and support to accelerate our transformations to a sustainable world' (FE, n.d. a).[3] Others are more suggestive of the 'rich tapestry' of a hybrid network, for example: 'at its core a

---

2 Interview and focus group participants comprised members of the Future Earth Science Committee, Interim Engagement Committee, Interim Secretariat, Alliance/Governing Council, and other actors involved in or aware of FE's development and implementation.
3 This has since been changed on the website to 'a major international research platform' – perhaps as an acknowledgement that there may be other such platforms.

"federation" of projects and other initiatives related to Global Environmental Change' (FE, n.d. c); an umbrella, tent, arena, network, and hub (e.g. interviews 7 and 8;[4] FE, n.d. d). Others still might fall into either, both or neither category, for example: 'first and foremost a community' (FE, n.d. e); part of a global innovation system (interview 9); amongst others.

Across the data, FE's imagined roles and functions are as diverse as its definitions and designations. They include (but are not limited to): integrating the existing GEC programmes; creating interdisciplinary, integrated, authoritative knowledge on the key challenges of GEC and sustainability (including climate change, biodiversity loss, and socio-economic change); informing policy, decision making and action in the UN system (such as the Sustainable Development Goals, the IPCC and the Framework Convention on Climate Change) and at other levels; raising, redirecting and coordinating research funds, activity and capacity; bringing about a change in the culture or practice of scientific research; involving existing and new communities of researchers, particularly from the social sciences, and engaging stakeholders; bringing together GEC and sustainability research communities and refocusing GEC research on sustainability challenges; all at local, national, regional, and/or international levels.

This wide-ranging diversity in defining what FE is and what its remit entails is acknowledged in passing in two early Q&A posts on the FE blog. In posing questions to the Chair of the FE Science Committee and to the Interim Director of FE, ICSU communications staff noted respectively that 'Future Earth seems to mean different things to different people' (Mengel, 2013) and 'Future Earth may seem like all things to all people' (Young, 2013). These statements reflect the roots of FE's plural definitions and purposes: on the one hand, many people have been involved (and have a stake) in its development, giving rise to a broad range of visions and interpretations; and on the other, the ambitions for its remit and scope are extremely broad even within single narratives of the initiative, incorporating a wide range of actors, knowledges, practices, phenomena, and scales.

4 In an effort to maintain anonymity, interview numbers (one per interview) have been attributed at random and do not reflect the order in which the interviews were conducted.

Although diverse understandings of what FE is and might do emerged from the interviews, there were varying degrees of awareness of and concern about these differences, with some interviewees retaining a strong sense of FE's purpose (from their perspective), and others viewing a lack of clarity around its identity, objectives and activities as potentially problematic. In mid-May 2014, one Interim Engagement Committee member argued for more specificity about FE's intended achievements, stating that 'Future Earth needs to really, really improve the definition of who and what it is and what its exact purpose is; I still think it's a bit too vague' (interview 10).

Lack of clarity or focus in FE's mandate has also been flagged as problematic by external stakeholders. In late May 2014, Ian Thornton, deputy director of the UK Collaborative on Development Sciences, blogged about FE, having attended an ICSU workshop. He suggests that despite considerable awareness of FE among UK stakeholders, there is confusion about 'what Future Earth will do, its value-add, and whether UK funders should engage more closely' (Thornton, 2014). He attributes this confusion to the challenges inherent in articulating what coordination of research can achieve, particularly in the context of an evolving organisational mandate, arguing that FE should pare back its goals and clarify what the secretariat will do: 'Is it mainly coordination? Or, evidence synthesis and policy influence? Or being a hub for debate?' (Thornton, 2014). Indeed, during the period of study, FE documents and actors suggested that the initiative (if not specifically the secretariat) is intended to take all three roles identified by Thornton, and more.

Overall, the analysis of the data suggests (at least) six linked points of tension or dilemmas between different visions and ambitions for FE, at times echoing the leviathan and hybrid-network models. Firstly, related to Thornton's point above, there is a potential conflict between visions of FE as a hub or arena for debate, dialogue and deliberation, and that of FE as a platform to deliver solutions-oriented, policy-relevant knowledge and innovation outputs. The ability to achieve both within the same institutional and conceptual framework has been questioned by scholars outside the initiative (Lövbrand et al., 2015).

This is closely linked to the second and third tensions. The second is between consensus and plurality, or between the desire to integrate knowledge (and also to create an integrated, authoritative, singular organisation, programme, or brand) and the inevitable multiplicity of a multi-scalar, multidisciplinary, multi-stakeholder international initiative and its forms of knowledge (Klenk and Meehan, 2015). The third tension is between the ambition to bring about new ways of doing science while also promoting and ensuring continued authority for existing ways. Here we might loosely align debate and plurality with the hybrid-network model, and solutions and singularity with leviathan (with caution, as STS has alerted us to the reductionism of binaries), but the new versus existing ways do not align as neatly with the models. Perhaps a hybrid network is more amenable to innovation or novelty, and the (authority of the) status quo more easily maintained and elevated by a leviathan.

One participant discussed FE's role as a forum for debate, highlighting tensions between new and old:

> To me, what's exciting about Future Earth is that it is pioneering a new approach to thinking about and doing science, which is integrated, interdisciplinary, co-produced, and it also should be a forum for encouraging debate and encouraging this approach but in a critically reflective way. [...] But I do think it faces real tensions because at the same time it's trying to give more prominence to some fairly standard ways of doing science.

The participant then identified a challenge:

> But the threat is that integrating everything becomes a kind of lowest common denominator and you lose, as it were, the cutting edge nature of that science – and I think both can happen there but I think we need to work very actively with those tensions and be aware of them and push on both fronts. (Interview 7)

The tension between singularity and multiplicity is perhaps heightened by the authority-building and fundraising aspirations for the initiative, particularly for those imagining its place in the UN system. One participant suggested that FE is 'not trying to be totalising', but that the ambition is to be 'the main place' and 'the natural platform' with which the UN, the European Commission and other international organisations wish to partner (interview 8).

Relating to authority building, there is a fourth tension between FE's ambition to be inclusive (of both existing and new research and stakeholder communities) versus upholding standards and setting limits on what counts as FE knowledge or approaches. On the one hand, interviewees spoke about working towards making FE 'a growing tangible concern that people understand and that they want to be part of' (interview 8), and opening up FE not only to the core projects of the existing GEC programmes but also to other science communities, including 'the mass of environmental social science across the world that by definition until now has been excluded from this dominant global environmental change community' (interview 7). However, they also spoke about the need to 'regulate the arena' (e.g. through peer review, critique and allowing the most robust claims to survive), whether because scholars and/or stakeholders are not 'fully aligned' with FE, or 'committed enough to the codes', or might 'clutter up the arena with false claims or with unfounded claims' (interview 8). Processes for affiliation with FE and ways of including other communities were being discussed and designed by the committees during the period these interviews were conducted, with varying ideas about the appropriate level of bureaucracy, who should be included and how.

Linked to this ambiguity around openness to participation or affiliation, there is a fifth tension between the directive action of establishing an initiative from the top (global and centralised) level down and the ambition to be responsive to the needs of society and to include bottom-up input from the scientific community. One participant argued that FE is 'a directive programme' and is not intended to do 'purely responsive science', that it needs to set 'broad strategic directions through the [strategic research agenda] process', but that there should be 'wide engagement' in putting those agendas together, and the research community should be given 'wide freedoms' to design research to meet them, without it being 'a completely blue skies programme' (interview 3).

The dilemmas between strategy/directiveness versus responsiveness, and between blue-sky research versus useful research, link to a sixth tension, which itself links back to new versus existing ways. This is the tension between a utilitarian focus on demand-driven science and the more traditional curiosity-driven model. Again, we might loosely align inclusivity and responsiveness with the hybrid-network model, and setting standards and strategy or direction with that of the leviathan. But, similar to the tension between new versus existing modes of science, utility versus demand and curiosity versus supply do not map neatly onto the models.

While thinking in terms of dilemmas might suggest mutual exclusivity, these tensions are not necessarily dichotomous. However, they are issues that those involved in FE were negotiating and navigating at the time of this study, and will probably continue to face as the initiative further develops. While this type of ambiguity can perhaps be expected in the early stages of a new initiative as its identity and remit are still in flux, it could be argued that some stabilisation may occur through the social order of recognised practices, structures and identities, particularly as the multiple roles and functions imagined for FE inform design decisions made about its governance, operations and activities.

For example, certain aspects may find some sort of stability in FE's organisational structure or its infrastructural architecture. In March 2016 the headline of the 'Get Involved' page on the FE website was changed from 'Future Earth is first and foremost a community' to 'Future Earth is first and foremost an open network committed to global sustainability.' This change was implemented in conjunction with the launch of the FE 'Open Network', an online networking tool to which anyone can sign up: (potentially) a concrete mechanism of openness and bottom-up input.[5] (Of course, the extent to which such mechanisms will be used or will enable collaboration on equal footing remains to be seen.)

Despite these points of stabilisation, as FE further develops and is rolled out and taken up at regional and national levels (whether through its regional hubs or centres and national committees, and/or by other initiatives affiliating, or in multiple other possible ways), the same and further ambiguities are bound to emerge. As the FE network extends and becomes more complex, the FE model is also extended, adapted and interpreted in a diverse range of contexts.

---

5 These developments were at least partly driven by personnel changes between the interim and permanent secretariat in 2015, and the associated shifts in perspectives and approaches.

Ambiguities in co-design and co-production

In addition to the tensions and ambiguities that emerge when thinking through and talking about FE's identity and functions, there is ambiguity around the concepts and roles of co-design, co-production and transdisciplinarity in FE. This ambiguity has also been regarded as problematic by some. As noted by science writer Jon Turney in an FE blog post (Turney, 2014), the *Design Report* (FE, 2013) and the 'Impact' section of the website (FE, n.d. b) perhaps prompt more questions than provide answers about these concepts. On the one hand, it is suggested that co-design and co-production require 'an active involvement of researchers and stakeholders during the entire research process' (FE, 2013: 22), but on the other, it seems that less active involvement of stakeholders may be preferred at some stages:

> Co-design and co-production of knowledge include various steps where both researchers and other stakeholders are involved but to different extents and with different responsibilities [...] Whilst researchers are responsible for the scientific methodologies, the definition of the research questions and the dissemination of results are done jointly. [...] It is also recognised that the focus for this way of working should be on where the research and stakeholder community feel that it will bring the greatest benefits. (FE, 2013: 23)

The website and *Design Report* do not address how to reconcile continuous engagement with the described division of responsibilities, nor how it should be decided where co-design and co-production bring the greatest benefits (and to whom). There are additional uncertainties on the roles of participants and stakeholders in co-design and co-production, the scale at which these endeavours should happen, and the anticipated outputs and outcomes, stemming in part from the differing rationales for undertaking such processes (and more broadly from differing imaginaries of science, society and the relationship between them) (Hadley Kershaw, forthcoming).

The predominant vision of co-design, co-production and transdisciplinarity in FE (in the documents and among many of the key actors) is that collaboration between researchers and stakeholders will ensure that the 'right' research questions are asked and therefore relevant or useful knowledge (and solutions) will be produced, and will be more likely to have impact as users have been involved in its production.

This utilitarian view corresponds to the logic of accountability as discussed on p. 111: redirecting research portfolios towards societal need (Lövbrand, 2011: 227), with an additional emphasis on impact of research. This type of rationale has been criticised for its normative assumption that co-produced and transdisciplinary research leads unproblematically to beneficial social solutions (Polk, 2014), and for restricting plurality, deliberation and social-science involvement (Lövbrand et al., 2015).

However, the interviews and focus groups conducted for this research revealed a broader range of understandings of co-design and co-production in FE. Some participants saw them as deliberative or reflexive processes in which multiple perspectives, commitments and knowledges are brought together, discussed and socially constructed:

> I would see the sort of co-construction agenda as about people, all these people, whether they're users, whether they're different scientists from different disciplines, coming to a situation bringing their own social commitments and drivers and understandings and assumptions about the world, and bringing those together and reflecting on them reflexively, realising that everybody's got a set of partial perspectives so those need to be deliberated on and debated out throughout the process, throughout the design, throughout the doing, throughout the communication. (Interview 7)

This view is more closely linked to visions of FE as an arena for deliberation, rather than as a platform for solutions, aligning with the logic of ontology described by Lövbrand (2011; following Barry et al., 2008) in which dominant frameworks and world views can be challenged, and ontological questions – including questions of common purpose – asked. That is not to say that participants voicing these opinions were not interested in finding solutions, having impact or 'making a difference'. However, this was more closely associated with extending agency and the rights of knowledge production or governance to non-academic actors, valuing their perspectives and knowledge, and democratising expertise:

> [Co-production] is to recognise that scientific knowledge or scientists are not the only people who hold relevant knowledge. That knowledge of practitioners, decision-makers, local communities, et cetera, is valid knowledge. […] That is why I like to call them knowledge partners,

because they are not just providers of additional data, they are not witnesses, they are active knowledge partners, their knowledge counts. (Interview 5)

A third view of co-design and co-production was also apparent: they were perceived by some as a threat. Many participants referred to the ideal of objective, pure science; either expressing concern about how to preserve it or noting that others in FE feel that it is challenged by co-design and/or co-production. Some of the most significant contestation arose around adherence or otherwise to this ideal, as well as around the notions of scientific independence and academic freedom. One participant suggested that the involvement of private-sector organisations (e.g. oil companies) in research could be dangerous, considering the elimination of conflicts of interest to be 'the hallmark of science':

> For me, in the middle [of the research process] there should be a complete separation, I do not believe that stakeholders should ever be involved in the process of generating data, interpreting data or peer reviewing data. That, to me, provides egregious conflicts of interests that ... will ruin the integrity of science for society, in my opinion. (Interview 2)

However, this participant was not totally against the co-design of research questions and priorities, as long as there was still support for curiosity-driven science beyond FE. Other participants felt that perceived risks of bias or conflict of interest associated with involving societal stakeholders were legitimate but manageable concerns. For example, one participant suggested that accountability could be built into the research process by stipulating that projects address questions of inequality when co-designing and/or co-producing with business or other powerful actors (focus group 2); another proposed that political co-option could be avoided by ensuring that the academic peer-review process would be undertaken by rigorous and sceptical 'top-rate' scientists (interview 8).

Some committee members noted that the ambiguity between these different understandings of co-design and co-production (and the different levels of commitment to requiring them as mandatory elements of any affiliated project) was potentially problematic. Several participants suggested that it would be necessary to develop common understandings of these concepts and common ground rules, principles, processes and practices in order to implement them and guard against conflict of interest and other threats. Others still stressed that there would probably always be multiple interpretations of these terms.

This latter view seems to strongly correlate with the hybrid-network/rich-tapestry model of FE, and raises the question of whether FE's institutional structure and accreditation/affiliation processes will be open and flexible enough to enable such multiple meanings and practices. Defined rules and shared understandings (temporary stabilisation) might be needed to avoid some of the tensions identified (such as conflict of interest, and ensuring that marginalised groups are considered when deciding with whom to co-produce) or to actually get something done (at least in a particular way, whether towards utility, democratisation, both or neither). However, any such rules or understandings may need to be settled and resettled as the initiative evolves – what one participant called an iterative 'learning process' (focus group 2).

So, where do these many differences, ambiguities and tensions leave us? Is FE best viewed as a leviathan, or a hybrid, or something else?

Living with ambiguity

The argument for FE as a leviathan might consider its emphasis on a new 'social contract' for science and society; the (potential) boundary building around what constitutes FE activities and knowledge; the singular nomenclature of 'Future Earth' (as opposed to, for example, 'Earth's Futures', or other plural alternatives); and the centralised, global coordination of the initiative. The argument for FE as a hybrid network might consider it as a boundary organisation (e.g. Miller, 2001) in its work to span science and politics; the (albeit initially imperfect (Padma, 2014)) regional distribution of the global secretariat hubs, regional centres, national committees and international projects, and the diversity of its global committee members; and its efforts towards inclusivity, such as the recent launch of the Open Network. Centralised documents and processes can be blunt instruments as they condense a broad range of views towards consensus or settlement, but perhaps it is unfair to look only at those instances of crystallisation in FE's development, when many other actors, knowledges, practices and phenomena operate within and around what some label 'Future Earth'.

For now it seems that there is something of both models in FE, depending on which lens is chosen, and perhaps it will continue to be an ambiguous double hybrid. However, analytically, it may be more useful to see FE as a continuously evolving network, assemblage or configuration – and part of bigger, broader networks – than as a rational, bounded entity (Pallett and Chilvers, 2015). It is then easier to take into account its multilayered development and iterations across different contexts, shaped by and shaping different actors, objects, cultures and practices; perhaps this view also opens up the possibility of taking a place within and/or shaping this initiative ourselves.

While FE faces many of the well-documented challenges of co-design, co-production and transdisciplinarity – and more, owing to the scale of its ambition – the various ambiguities that emerge in thinking, talking about and practicing FE, and what co-design and co-production might mean within it, are not necessarily problematic. They are inevitable and even necessary. They could keep FE, and especially co-production, alive.

Ambiguity can be seen as flexibility and openness, making space for conversation and the negotiation of meaning (Nerlich and Clarke, 2001). This is particularly the case when FE is imagined as an ongoing experiment, which may never reach a conclusion about how this type of work 'should' be done – indeed, in which such conclusions would be undesirable as they would close down the potential to adapt to new circumstances, contexts and actors. FE is part of broader systems, networks and 'ecologies of participation', rather than a dislocated, self-contained institution or event to be evaluated for its successes and failures against rigid pre-given standards or norms (Chilvers and Kearnes, 2016; Irwin, 2006). The notion that FE is stepping into the unknown and 'feeling the way', and that co-production is something that can only be worked out in practice (at 'field level') in specific projects and contexts, has been acknowledged by various FE actors, whether seeing co-production as a 'messy social experiment' (Turney, 2014), or thinking about a White Paper on engagement as 'a testable hypothesis' (interview 9).

From this perspective, FE should be treated with care and analytical generosity. Although its challenges and limitations should be acknowledged, it is also constructive to recognise that, during this study at least, FE had already introduced changes in comparison with the existing GEC programmes. Simply starting conversations about why and how to co-produce research has the potential to introduce unfamiliar questions and perhaps reflexivity to existing projects, and institutional developments – such as the introduction of an engagement committee in addition to the standard scientific committee, and thematic rather than disciplinary organisation of research communities – have the potential to effect further change to well-established social and epistemic orders:

> I mean, one really good thing is even having discussions about [power relations], because in somewhere like the IGBP Science Committee I can assure you the notion of talking about power would never have come up [laughs]. (Interview 9)

Perhaps it is productive to ask how FE can maintain experimentality while also putting policy and structures in place. How can it – and similar programmes – achieve a balance between bringing itself into being and allowing space for development, new and multiple perspectives, and bottom-up initiative? (How) can intangible processes and outcomes of research be valued in a culture of solutions orientation, policy relevance, and increasing quantification and audit (Strathern, 2000)? How can tensions between openness and closure be negotiated?

Other scholars have suggested that humility, institutional reflexivity and organisational learning are key to ensuring ongoing space for plurality, flexibility, diversity and capacity building (Beck et al., 2014; Felt and Wynne, 2007; Jasanoff, 2003; Pallett and Chilvers, 2015). There are indications that FE aspires to achieve this type of practice, whether through a reflexive form of co-production, through monitoring and evaluating its own processes and outcomes, or in acknowledging that FE is an ongoing learning process. Future research could (carefully and generously) explore to what extent – and how – FE is managing to achieve this within and beyond its ever-expanding and complex structure and networks.

Acknowledgements

Many thanks are due to Future Earth and all of the research participants who generously gave their time and input. Sincere thanks as well to Brigitte Nerlich and Sujatha Raman for their invaluable guidance, support and feedback on this chapter and the broader PhD project, and to Brigitte for the chapter title, after Shapin and Schaffer (1985).

References

Barron, E. J. (1994). IGBP core projects: A 'rich tapestry' or 'flagship' model. *Global Change Newsletter*, 17, 2. Retrieved 15 March 2016 from: http://igbp.net/download/18.950c2fa1495db7081e196/1417001148822/NL_171994.pdf.

Barry, A., Born, G., and Weszkainys, G. (2008). Logics of interdisciplinarity. *Economy and Society*, 37(1), 20–49.

Beck, S., Borie, M., Chilvers, J., Esguerra, A., Heubach, K., Hulme, M., Lidskog, R., et al. (2014). Towards a reflexive turn in the governance of global environmental expertise: The cases of the IPCC and the IPBES. *GAIA-Ecological Perspectives for Science and Society*, 23(2), 80–87.

Beck, S., and Forsyth, T. (2015). Co-production and democratizing global environmental expertise. In S. Hilgartner, C. A. Miller and R. Hagendijk (eds), *Science and Democracy: Making Knowledge and Making Power in the Biosciences and Beyond* (pp. 113–132). New York: Routledge.

Belmont Forum (2011). The Belmont Challenge: A Global, Environmental Research Mission for Sustainability. *Belmont Forum*. Retrieved 25 March 2016 from: https://belmontforum.org/sites/default/files/documents/belmont-challenge-white-paper.pdf. [Now available at: http://igfagcr.org/sites/default/files/documents/belmont-challenge-white-paper.pdf.]

Bensaud Vincent, B. (2014). The politics of buzzwords at the interface of technoscience, market and society: The case of 'public engagement in science'. *Public Understanding of Science*, 23(3), 238–253.

Bovaird, T. (2007). Beyond engagement and participation: User and community coproduction of public services. *Public Administration Review*, 67(5), 846–860.

Campbell, H., and Vanderhoven, D. (2016). Knowledge that matters: Realising the potential of co-production. N8/ESRC Research Programme. Retrieved 6 June 2016 from: www.n8research.org.uk/view/5163/Final-Report-Co-Production-2016-01-20.pdf. [Now available at: www.n8research.org.uk/media/Final-Report-Co-Production-2016-01-20.pdf.]

Chilvers, J., and Kearnes, M. (2016). Science, democracy and emergent publics. In J. Chilvers and M. Kearnes (eds), *Remaking Participation: Science, Environment and Emergent Publics* (pp. 1–27). London and New York: Routledge.

Durose, C., and Richardson, L. (2015). *Designing Public Policy for Co-Production: Theory, Practice and Change*. Bristol: Policy Press.

FE (2013). *Future Earth Initial Design: Report of the Transition Team*. Paris: ICSU. Retrieved 25 March 2016 from: www.futureearth.org/sites/default/files/Future-Earth-Design-Report_web.pdf.

FE (2014). *Future Earth 2025 Vision*. Paris: ICSU. Retrieved 25 March 2016 from: www.futureearth.org/sites/default/files/files/Future-Earth_10-year-vision_web.pdf.

FE (n.d. a). Who we are. *Futureearth.org*. Retrieved 25 March 2016 from: http://futureearth.org/who-we-are. [Earlier version (accessed April 2015) now available at: http://web.archive.org/web/20150430035528/http://futureearth.org/who-we-are.]

FE (n.d. b). Impact. *Futureearth.org*. Retrieved 25 March 2016 from: http://futureearth.org/impact.

FE (n.d. c). Research. *Futureearth.org*. Retrieved 25 March 2016 from: http://futureearth.org/projects. [Now available at: http://web.archive.org/web/20160329130624/http://futureearth.org/projects.]

FE (n.d. d). About Future Earth. *Futureearth.org*. Retrieved 2 March 2014 from: http://futureearth.info/blog/about-us. [Now available at: https://web.archive.org/web/20140227152100/futureearth.info/about-us.]

FE (n.d. e). Get involved. *Futureearth.org*. Retrieved 24 February 2016 from: www.futureearth.org/get-involved. [Now available at: http://web.archive.org/web/20160224234058/http:/www.futureearth.org/get-involved.]

Felt, U., and Wynne, B. (2007). *Taking European Knowledge Society Seriously: Report of the Expert Group on Science and Governance to the Science, Economy and Society Directorate*. Brussels: Directorate-General for Research.

Funtowicz, S., and Ravetz, J. (1993). Science for the post-normal age. *Futures*, 25(7), 739–755.

Gibbons, M., Limoges, C., Nowotny, H., Schwartzman, S., Scott, P., and Trow, M. (1994). *The New Production of Knowledge: The Dynamics of Science and Research in Contemporary Societies*. London: Sage.

Godin, B. (1998). Writing performative history: The new Atlantis? *Social Studies of Science*, 28(3), 465–485.

Hadley Kershaw, E. (forthcoming). Co-producing Future Earth: Ambiguity and experimentation in the governance of global environmental change research. PhD thesis, University of Nottingham.

Haraway, D. (1991). *Simians, Cyborgs and Women: The Reinvention of Nature*. New York: Routledge.

Hessels, L.K., and Van Lente, H. (2008). Re-thinking new knowledge production: A literature review and a research agenda. *Research Policy*, 37, 740–760.

ICSU (2009). *Developing a New Vision and Strategic Framework for Earth System Research*. ICSU Visioning Process Paper. Paris: ICSU.

ICSU (2011). *Summary of the 3rd Earth System Visioning Meeting*. ICSU Meeting Report. Paris: ICSU.

ICSU (n.d.). Future Earth. *ICSU.org*. Retrieved 25 March 2016 from: www.icsu.org/future-earth. [Now available at: http://web.archive.org/web/20170416140248/www.icsu.org/future-earth.]

Irwin, A. (2006). The politics of talk: Coming to terms with the 'new' scientific governance. *Social Studies of Science*, 36(2), 299–320.

Jasanoff, S. (2003). Technologies of humility: Citizen participation in governing science. *Minerva*, 41(3), 223–244.

Jasanoff, S. (2004). The idiom of co-production. In S. Jasanoff (ed.), *States of Knowledge: The Co-Production of Science and Social Order* (pp. 1–12). London: Routledge.

Jasanoff, S., and Wynne, B. (1998). Science and decision-making. In S. Rayner and E. Malone (eds), *Human Choice and Climate Change. Volume 1: The Societal Framework* (pp. 1–88). Columbus, OH: Batelle Press.

Klenk, N., and Meehan, K. (2015). Climate change and transdisciplinary science: Problematizing the integration imperative. *Environmental Science and Policy*, 54, 160–167.

Lahsen, M. (2016). Toward a sustainable Future Earth: Challenges for a research agenda. *Science, Technology and Human Values*, 41(5), 876–898, doi: 10.1177/0162243916639728.

Latour, B. (2011). Love your monsters: Why we must care for our technologies as we do our children. *Breakthrough Journal*, 2, 21–28.

Law, J. (1991). Introduction: Monsters, machines and sociotechnical relations. In J. Law. (ed.), *A Sociology of Monsters: Essays on Power, Technology and Domination* (pp. 1–25). London: Routledge.

Lövbrand, E. (2011). Co-producing European climate science and policy: A cautionary note on the making of useful knowledge. *Science and Public Policy*, 38(3), 225–236.

Lövbrand, E., Beck, S., Chilvers, J., Forsyth, T., Hedrén, J., Hulme, M., Lidskog, R., et al. (2015). Who speaks for the future of Earth? How critical social science can extend the conversation on the Anthropocene. *Global Environmental Change*, 32, 211–218.

Mengel, J. (2013). Q&A with Mark Stafford Smith, Science Committee Chair. *Future Earth Blog*, 30 July. Retrieved 25 March 2016 from: http://futureearth.org/blog/2013-jul-30/qa-mark-stafford-smith-science-committee-chair.

Miller, C. A. (2001). Hybrid management: Boundary organizations, science policy, and environmental governance in the climate regime. *Science, Technology and Human Values*, 26(4), 478–500.

Nerlich, B., and Clarke, D. D. (2001). Ambiguities we live by: Towards a pragmatics of polysemy. *Journal of Pragmatics*, 33(1), 1–20.

Padma, T. V. (2014). Future Earth's 'global' secretariat under fire. *SciDevNet*, 23 July. Retrieved 31 May 2016 from: www.scidev.net/global/sustainability/news/future-earth-global-secretariat.html.

Pallett, H., and Chilvers, J. (2015). Organizations in the making: Learning and intervening at the science–policy interface. *Progress in Human Geography*, 39(2), 146–166.

Pohl, C., Rist, S., Zimmermann, A., Fry, P., Gurung, G. S., Schneider, F., Speranza, C. I., et al. (2010). Researchers' roles in knowledge co-production: Experience from sustainability research in Kenya, Switzerland, Bolivia and Nepal. *Science and Public Policy*, 37(4), 267–281.

Polk, M. (2014). Achieving the promise of transdisciplinarity: A critical exploration of the relationship between transdisciplinary research and societal problem solving. *Sustainability Science*, 9(4), 439–451.

Ribeiro, B. E., Smith, R. J. D., and Millar, K. (2016). A mobilising concept? Unpacking academic representations of responsible research and innovation. *Science and Engineering Ethics*, 23(1), 81–103 [epub ahead of print]. Retrieved 10 June 2016 from: http://link.springer.com/article/10.1007/s11948-016-9761-6.

Robinson, J., and Tansey, J. (2006). Co-production, emergent properties and strong interactive social research: The Georgia Basin futures project. *Science and Public Policy*, 33(2), 151–160.

Shapin, S., and Schaffer, S. (1985). *Leviathan and the Air-Pump: Hobbes, Boyle, and the Experimental Life*. Princeton, NJ: Princeton University Press.

Star, S. L., and Griesemer, J. R. (1989). Institutional ecology, translations and boundary objects: Amateurs and professionals in Berkeley's Museum of Vertebrate Zoology, 1907–39. *Social Studies of Science*, 19(3), 387–420.

Stirling, A. (2008). 'Opening up' and 'closing down': Power, participation, and pluralism in the social appraisal of technology. *Science, Technology, and Human Values*, 33(2), 262–294.

Strathern, M. (ed.) (2000). *Audit Cultures: Anthropological Studies in Accountability, Ethics, and the Academy*. London and New York: Routledge.

Thornton, I. (2014). What does coordination actually achieve? Some thoughts on Future Earth. *UKCDS Blog*, 29 May. Retrieved 25 March 2016 from: www.ukcds.org.uk/blog/what-does-coordination-actually-achieve-some-thoughts-on-future-earth.

Turney, J. (2014). Contemplating co-production. *Future Earth Blog*, 23 July. Retrieved 25 March 2016 from: http://futureearth.org/blog/2014-jul-23/contemplating-co-production.

van der Hel, S. (2016). New science for global sustainability? The institutionalisation of knowledge co-production in Future Earth. *Environmental Science and Policy*, 61, 165–175.

Young, D. (2013). Q&A with Frans Berkhout, Future Earth Interim Director. *Future Earth Blog*, 15 July. Retrieved 25 March 2016 from: http://futureearth.org/blog/2013-jul-15/qa-frans-berkhout-future-earth-interim-director.

Ziman, J. (2000). *Real Science: What It Is, and What It Means*. Cambridge: Cambridge University Press.

7

'Opening up' energy transitions research for development

Alison Mohr

The term 'energy transition' has gained increasing traction internationally in research and policy communities seeking tools and concepts to study and explain the transformation to more sustainable energy systems. A significant limitation of the energy transitions literature is that much of it relates to the experiences of industrialised countries in the global North attempting to transition to sustainable energy futures. Yet there is also an urgent need to understand the potential nature of emergent transitions in the industrialising countries of the global South where energy is inextricably linked to development, but also where decarbonising the energy system may not be the core or most pressing objective. This prompts the question of how well transitions concepts and frameworks, shaped largely by the experience of research based in industrialised countries, translate when applied in developing contexts. In this respect, transitions are relatively unchartered territories that hide numerous monsters, not least in the contested nature of the term itself. Understandings and expectations of energy transitions may differ across the North–South divide; thus, how transition is framed and by whom will have material implications for livelihoods, raising the spectre of social justice and other monsters in energy transitions research for development.

This chapter is concerned with opening up research, specifically the case of global-North-led research on sustainable energy transitions in the global South, to examine underlying assumptions, blind spots and gaps, but also to identify alternative ways of doing transitions research that are more responsive to the public and social priorities of the transitioning communities. Alternative or competing priorities for transitions research are best detected through a process of co-design; in this case the co-design of community energy systems on the basis of bottom-up iterative public and community engagement, as one of these alternative ways of doing transitions research. Making transitions research 'open' from the bottom up presupposes the embedding of democratic values such as inclusiveness and social justice.

I will explore how attempts to open up transitions research through a process of co-design can facilitate or impede values of inclusiveness and social justice by drawing on the experience of the ongoing Solar Nano-Grids (SONG) project, which aims to explore ways of co-designing small-scale off-grid energy systems in rural Kenya and Bangladesh through community engagement. The SONG project emerged as a response to the perceived shortcomings in these countries of solar-home-systems (SHS) installations that provide individual households only with electricity for lighting and small appliances. While households undoubtedly gain developmental benefits from access to SHS, the degree to which they address the poverty of household members and their ability to generate an income is far less clear. In response, SONG systems aim to provide additional collective community benefits by the provision of excess energy for small-scale economic activities using, for example, posho (maize meal) mills and egg incubators.

SONG is an expressly transdisciplinary project that brings together social scientists, engineers and not-for-profit renewable-energy organisations engaged in the co-design of all aspects of the energy-transitions research, from community engagement and implementation through to monitoring and evaluation. The role of social science in SONG is distinct in a number of ways. By research, I mean social science as well as engineering research. Many of the chapters in this book are about research understood as 'scientific research', which social scientists are critically investigating from the outside, or post hoc (of which ELSI, research on ethical, legal and social implications, that arose as an extramural research programme to the Human Genome project, is a prime example). In SONG, social scientists are engaged in problem solving, not just problematising from afar, and are actively involved in the design, implementation and redesign of socio-technical systems.

SONG's approach to energy transitions research is novel in that it is taking transitions theory, and in particular the concept of niche experimentation, and applying it in practice to community niches in Kenya and Bangladesh to facilitate energy transitions. But the application of transitions concepts and tools in developing contexts is not without its practical challenges. What are the limits of research frameworks for energy transitions when they are transferred from the global North to the global South? How does co-design of energy-transitions research actually work in practice? Who and what does it need to take into account?

The chapter proceeds with a brief introduction to the transitions literature through the lens of its principal theoretical frameworks. It outlines key concepts and identifies notable limitations, including the limited but emerging application of transitions research in developing contexts. Taking the position that the absent or naive conceptualisations of the dominant frameworks stem from the neglect of issues of distributive and procedural justice, the remaining sections show how the co-design of the SONG project was fundamental to addressing these justice deficits. The chapter concludes by reflecting on implications for energy transitions research arising from applying justice principles in practice.

Limitations of transitions research frameworks and concepts

The term 'transition' emerged from a body of literature on socio-technical transitions (Geels, 2005; Geels and Schot, 2007) that refers to system-wide changes involving the complex reconfigurations of technology, policy, markets, material infrastructure, scientific knowledge, and social and cultural practices over time towards more sustainable systems, in areas such as energy (Newell and Mulvaney, 2013). Systems-level innovations occur through 'the interplay of dynamics at multiple levels' (Geels, 2005: 368). A micro-level niche is nested within and interacts with a patchwork of meso-level regimes and macro-level landscape developments to form a socio-technical system. The core notion of this multi-level perspective framework is that the increasing traction of niche innovations from below and pressure exerted from above by changes in external factors at the landscape level serve to destabilise the extant socio-technical regime, thus creating opportunities for niche innovations to become systemically embedded.

Of particular relevance to the focus of this chapter on the community niche as an innovation space for the development of off-grid energy systems is strategic niche management, a subcategory of the multi-level perspective. A niche refers to a protected space or 'incubation room' (Geels and Schot, 2007) at the micro level, where 'radical innovations' (Geels, 2011) are nurtured away from the normal pressures of market forces and technical performance to enable essential learning to take place (Smith et al., 2014). The niche performs an important function in propagating transformative innovations and providing a footing for their future establishment in the dominant socio-technical regime. A key aspect of the strategic-niche management framework is that it directs attention to the co-evolution of stakeholders' visions of particular innovation-enabled futures, the learning gained from experimenting in everyday settings and the societal embedding of new socio-technical practices relevant to that particular innovation (Ockwell and Byrne, 2015).

The design and implementation of socio-technical transitions projects, however, have too often been narrowly framed as linear, top-down processes driven by the techno-economic priorities of implementers that limit the ability to provide just and equitable solutions responsive to the priorities of the communities in transition. This has led to an increasing recognition of the need to address the justice deficits of transitions processes. The following paragraphs identify two prominent deficits in the socio-technical transitions literature – a lack of consideration of the politics of framing and of socio-spatial relations and dynamics – that must be attended to if transitions are to evolve beyond their apolitical and technocratic origins to become more socially just (Goldthau and Sovacool, 2012).

Politics of framing

The act of transitioning from one socio-technical system to another is inherently political, yet existing analyses do not adequately address the politics of transitions (Lehtonen and Kern, 2009; Meadowcroft, 2009; Smith and Stirling, 2007). Alongside theories about the multi-level dynamics of system innovations, socio-technical transitions research has developed prescriptive formulations for the management of transitions towards more sustainable ends. Such forms of transitions management have traditionally been led by policy and stakeholder elites, who have been criticised for their tendency toward techno-economic determinism and their lack of critical reflexivity regarding who participates in, and the politics of, transitions by management (Lawhon and Murphy, 2011). Reflexive questions such as 'Whose transitions?' and 'A transition from what to where?' aim to reveal the politics of whose priorities should steer the transitions agenda.

In any particular energy transition, the matter of who gets to define what transition means will have a direct bearing over the outcomes of the transition. Is it a transition from traditional to modern sources of energy, or from centralised to distributed energy systems and governance (or vice versa), or from no or low access to energy to secure and reliable access? For energy-impoverished rural communities in developing countries a transition may not be about renewable energy but rather the more fundamental issue of access to energy, whatever its source. A key point is that it depends on who is allowed to define the agenda. So the issue of whether a particular transition should aim to mitigate energy poverty, broadly speaking, or to increase access to energy, or even to reduce carbon emissions, is essentially an issue of the politics of framing and the radically different political understandings and actions that underpin these claims. How a transition is framed, and by whom, may have material implications for those whose livelihoods will be improved or impoverished.

In an attempt to open up the politics of transitions, Stirling (2014) draws a stark distinction between singular linear transitions driven by technological innovation and progressive transformations involving social as well as technological innovations shaped by a plurality of values and visions. Stirling argues persuasively that truly transformative and sustainable systemic change emerges out of the collective, bottom-up, socio-culturally embedded energy knowledge and practices of those in transition.

Socio-spatial relations and dynamics

The socio-technical transitions literature is also notable for a relative absence of reflection about the specific spatial contexts and conditions in which empirical case studies are situated, not least from a comparative perspective. Coenen et al. (2012: 969) claim that socio-technical transitions processes are characterised by 'missing or naïve conceptualisations of space', and that such 'analyses have overlooked where transitions take place, and the socio-spatial relations and dynamics within which transitions evolve'. A lack of sensitivity to specific sociospatial dynamics of transitions means that the intrinsic diversity arising from the variety of local, social, cultural and institutional conditions, and stakeholder priorities and resources across those spaces has not been adequately acknowledged or taken into consideration. Coenen et al. make the point that transitions researchers would do well to take a closer look at the spatial unevenness and inequalities of transition processes whether it be from the perspective of global research and policy networks or of local community niches.

Recently, transitions scholars have begun to address the lack of attention to the role of space (see Raven et al., 2012; Smith et al., 2010) by paying particular attention to the places in which niches are created, their organisation and scale of production, and how these places are shaped by relations between and among key stakeholders. Attention to space and scale is vital, given their potential importance when analysing energy transitions that span local and global contexts, such as the creation of global networks in which poorer parts of the global South provide resources for the energy ambitions of the richer North (van der Horst and Vermeylen, 2011), or where, in a globalised market for limited resources, they face the adverse knock-on effects of global-North investments in energy production (Raman and Mohr, 2013). Geels (2010), too, recognises the theoretical limitations of the existing literature for dealing with the additional complexities of emergent or future sustainability transitions, including the uncertainty involved in anticipating their potential impacts across time and space.

In this chapter, the off-grid nature of SONG systems necessitates a specific local perspective of the socio-spatial dynamics of communityenergy-transitions niches rather than the local global perspective of niche development trajectories used by Geels and Raven (2006) in the multi-level perspective to describe how local niche experiments can be scaled up as cumulative local outcomes reveal generic lessons and rules for widespread adoption. While recognising that the local is always situated in broader networks that operate at different scales and that a purely local-scale transition analysis risks ignoring the fact that localised activities and resources are subject to external pressures, a broad local–global perspective is beyond the scope of this current chapter.

Bridging the justice divide in North–South energy transitions research

The socio-technical transitions approach has in recent years begun to be applied in developing countries. A special edition of *Environmental Science and Policy* introduced by Berkhout et al. (2010) applied these ideas to developing Asia, to see what they revealed about the wider impacts of socio-technical innovations developed via niche experiments. A similar case study on technology innovation and transfer focuses on China (Wang and Watson 2010). Work on energy transitions in sub-Saharan Africa is limited, but researchers from the University of Sussex have published a number of studies focusing on SHS in Kenya. Byrne (2011) has shown how attempts to replicate the success of a Kenyan SHS initiative in neighbouring Tanzania failed as insufficient attention was paid to the specificities of the local context. A subsequent related study by Byrne et al. (2014) further demonstrates that context matters, as innovation processes are shaped as much by political, social and environmental forces as by powerful economic and institutional interests. Being aware of context is therefore critical for the replication of transitions initiatives across neighbouring territorial borders. Moreover, given its limited application in developing countries, it is reasonable to question the relevance of the literature framed by research agendas in the global North to the direct insertion of the findings into global South contexts. Assumptions underpinning global North research agendas, including those about universal access to electricity, the nature and pace of change between energy systems, and the priorities for systemic change, might not directly translate in developing contexts.

The absence of attention to the political and socio-spatial dimensions within the dominant frameworks for conceptualising sustainable-energy transitions may stem from a neglect of issues of distributive and procedural justice. Eames and Hunt (2013: 47) reason that 'Given the fundamental role that energy and energy technologies play in restructuring socio-economic and socio-ecological relations it is perhaps surprising that greater attention has not previously been paid to exploring the equity and justice implications of energy systems transitions.' When researchers from the global North intervene in sustainable-energy transitions for socio-economic development and improving livelihoods in energy-impoverished rural communities in the global South, questions of equity and justice take on a more critical edge. But is providing universal access to clean, affordable and reliable energy enough to resolve injustices in energy systems transitions? Pulido (1996) suggests that changes to power relations, cultural practices and systems of meaning are also required, beyond mere (socio-) economic restructuring and redistribution. This hints at the inadequacy of principles of distributive justice by themselves to acknowledge and address the range of valid issues of justice. A contemporary of Pulido, Lake (1996) agrees that it is impossible to have thorough distributive justice without justice in the procedures for producing that distribution.

In North–South collaborations it is imperative, therefore, that the key justice deficits of the dominant theoretical frameworks in relation to both the distributive and procedural dimensions of transitions research and practice should be addressed. Drawing on Schlosberg's (2007) work on defining what environmental justice means, this chapter proceeds with the argument that the theory and practice of justice necessarily include distributive justice (to address socio-spatial inequalities), but must also embrace notions of justice based in a recognition of differences and broad participation in the politics of transitions in the form of procedural opportunities. Accordingly, distributive justice refers to how the spatio-temporal distribution of risks and benefits is felt across social groups and communities and within and across generations, while procedural justice relates to who is granted both access to and participation in decision-making processes, as well as whose and what knowledge, interests and values are afforded recognition in such processes (Schlosberg, 2007).

Drawing on the specific case of SONG systems implementation for socio-economic development and livelihoods enhancement in two community niches – Lemolo B and Echareria – in Nakuru County, Kenya, the sections below detail instances where principles of distributive and procedural justice were applied in practice to inform the co-design of energy transitions in these communities. It is important to note, as observed by Lake (1996), that principles of distributive and procedural justice are mutually interdependent.

The co-design of just energy transitions in Kenya

Distributively just transitions

Fundamentally, a just transition requires attention to how the socio-spatial organisation of community niches affects community cohesion and the capacity of individuals to decide on a common energy future. To this end, the SONG project aimed to alleviate existing socio-spatial inequalities and prevent new ones, driven by factors such as age, gender, ethnicity and income, through an inclusive process of co-design.

The highly dispersed nature of the dwellings in the two villages selected by the Kenyan partner necessitated a redesign of the nano-grid systems to enable the distribution of power by portable batteries rather than lines, to accommodate the long distances between the solar array and central charging hub and individual dwellings. Redesigning the system this way meant that access was not limited to just those few households in a required radius, but also meant that the siting of the solar array was determined more by what would be a suitably secure and accessible location for the whole community than by who would benefit most from having the array sited nearby. Such decisions are highly political, even within a relatively small community niche. In the case of Echareria, the village consists of three spatially, socio-economically and tribally disparate settlements of internally displaced people. The settlements correspond to waves of intercommunal conflict linked to electoral violence. The first settlement was established in 1992, and households were given large blocks of land at the bottom of a hill, while the second and third settlements were established in 2001 and 2007–2008, respectively, and located progressively up the hill on decreasingly smaller blocks of land. The SONG project team, sensitive to the socio-economic and cultural asymmetries in the community, was therefore mindful not to create or exacerbate any tensions between these different groups by ensuring that physical and financial access to the system was as equitable as possible.

Women and children tend to be disproportionately affected by adverse health and environmental impacts caused by toxic smoke and fumes from 'unclean' energy as energy-related air pollution occurs mainly in and around households, particularly in developing countries. The major energy sources in both communities were kerosene for lighting and wood for cooking and heating. Replacing kerosene with solar electricity for lighting will improve air quality and help to reduce eye and respiratory infections, which are listed among the top ten diseases at the health clinic located nearest to Lemolo B. While electric cookers are beyond the load capacity of the SONG system, the women and children of the two communities stand to benefit from the transition to solar electricity in other ways, including improved educational achievement of children relative to those from non-electrified households; enhanced livelihood opportunities for women, leading to their improved economic status and self-reliance; and increased leisure time. Such claims support the point of Smith et al. (2010) that sustainable-energy transitions are more about distributed social mobilisation than technological innovation. Attention to differences across the full spectrum of intragenerational relationships to energy in the community niche is therefore important so that any concerns about distributive justice might be investigated within this ' "[micro]geography" of beneficiaries and risk-bearers' (Eames and Hunt, 2013: 58).

Finally, the anticipated human and environmental harms to future generations caused by anthropogenic climate change are widely acknowledged and debated in the climate justice literature (Okereke, 2010). Yet the intergenerational injustices, or opportunities, arising from sustainable-energy transitions are less well studied but equally deserving of attention, particularly as today's youth and subsequent generations will bear the consequences of transitions already underway. The aspirations of youth and future generations in terms of the role of energy in socio-economic development and livelihoods enhancement are likely to differ substantially from those held by their parents or village elders. This justice dilemma highlights the difficulty of taking future generations into account in current transitions, while also raising questions about what are the best research processes to use to deliver procedural justice.

Newell and Mulvaney (2013) make the related point that, to avoid reproducing or exacerbating existing intra- and intergenerational environmental inequalities in terms of exposure to ill health and localised degradation linked to global environmental threats, as discussed above, transitions also need to be environmentally just.

Procedurally just transitions

If communities provide a direct source of information on the effects of transitions on their social relations, cultural practices and livelihoods (and vice versa), how can transitions research be procedurally just if participation is contingent solely on the basis of having socio-technical expertise? Drawing parallels with Scharpf's (1999) thesis that democratic legitimacy refers to both the inputs and the outputs of a political system and that the two dimensions are mutually reinforcing, it can be assumed that the absence of attention to principles of justice in transitions would limit the public value of such processes. Thus, for any profound and lasting system change to occur, it is vital for households and communities to become co-designers and co-producers of their own energy transformation. By making little or no provision for communities to become informed about and included in the process of transitions research, implementation and governance, transformative change is impeded. Merely implementing transitions research agendas in the interest of the public good is not enough; the extent to which the justice dimensions of transitions research are recognised and attended to will have a direct bearing on its effectiveness in terms of creating public value for the communities involved.

A major limitation of previous SHS implementation in Kenya by development agencies and researchers has been a lack of routine engagement with project communities about their socio-economic priorities and aspirations, and the role of energy in achieving these. Where this has been the case, the public value of transitions research was not co-produced, rather it was imposed in a top-down fashion and the views of those who were to use the services were not adequately taken into account. Legitimate public value requires service providers to be more responsive to the needs of service users. The SONG project aimed to remedy this oversight by using the communities' views and values to inform the goals of the research and the design of the solar-energy systems. To this end, the project team recognised the need for more direct and innovative opportunities for the two communities to participate in the script of community energy services and technologies. Transitioning communities should not only receive the energy services they need, but also play a role as co-designers in the shaping and delivery of those services.

This challenge is influencing the way the SONG project is approaching the design and implementation of the SONG systems, with an emphasis on mutual learning and iterative development and deployment. To engage the transitioning communities in the co-design of their own energy futures, the project team embarked upon an in-depth community-engagement process with the twin aims of assessing the energy needs and aspirations at the level of the household, and assessing the socio-economic priorities and aspirations of the wider community. We adopted a mixed-methods approach to inform the design and development of SONG systems tailored to the particular contexts of each of the communities, including: a physical survey to map the spatial distribution of households, existing services, natural resources and proximity to external services; a household survey to identify the socio-demographic make-up of each and their livelihoods, the range, cost, and purpose of fuels and appliances currently used and expenditure on each; and semi-structured interviews with county- and national-level stakeholders to understand the broader socio-political barriers to and opportunities for electricity access for rural communities. The combined findings and continuity of contact helped us to gain a comprehensive understanding of the local niche contexts and to identify key issues for in-depth exploration in subsequent focus groups.

When considering how to structure our focus groups we were mindful of the potential distributive intra- and intergenerational injustices in each of our communities, as discussed above. But intra- and intergenerational justices also have a procedural dimension. We accordingly organised our focus groups into three broadly representative social groups – elders, women and youth – with the assumption that each broad category would have a different relationship to energy provision and use and therefore differing expectations of transition based on (among other things) their status, gender and age. For clarification, the gender balance among the elders was heavily biased towards men, while the gender split among the youth, defined in Kenya as between the ages of 18 and 30, was only slightly biased towards men.

A transition to a sustainable future and the deep structural changes that it entails places responsibility on communities to work cooperatively and inclusively to make decisions. However, this expectation may place a burden on communities if they lack cohesion. As Fuller and Bulkeley (2013: 68) note, 'area-based communities do not immediately imply the ready existence of a "community", there may be multiple overlapping communities of interest'. The notion of a community is a relatively recent and somewhat artificial construct in the case of Echareria, which consists of sequential settlements of internally displaced people from different tribal groups and with decreasing socio-economic means. Thus, enacting a sustainable-energy transition in this community niche not only raises critical issues of distributive justice in terms of the ways in which risks and benefits are allocated, but equally critical issues of procedural justice, depending on whether the transition can address more fundamental issues of inequality in terms of members' recognition and participation in community decision making. Including the knowledge, interests and values of participants from all three settlements was vital to achieving a balanced set of views in the household survey and focus groups.

Issues of distributive and procedural justice in North–South energy transitions collaborations do not cease to be important upon the physical implementation of the transformative technology. Pulido (1996) claims that injustices cannot be resolved solely through (socio-)economic restructuring or redistribution, but also require changes to existing social relations, cultural practices and systems of meanings. This still holds if energy transitions are to be truly transformative. In Kenya, the SONG team witnessed first-hand the continuing control and maintenance of one mini-grid system by its project implementers, who are based in the global North. This approach to governing is the antithesis of a just transition. The SONG project has taken the opposite route. It has worked with existing leaders and governance structures to facilitate the setting up of broadly inclusive and formally constituted village energy committees (VECs) in each of the communities. The role of the VECs is in part to ensure the project has a social value to the community as a whole by running a number of small enterprises such as egg incubators or posho mills, where the income is used to further expand the range of community services, to support individual entrepreneurialism or to extend to poorer members of the community access to energy services. The VECs will also have the responsibility to implement maintenance and repair training programmes for local youth for the better management of the systems over their lifetime.

Conclusions

This chapter was concerned with opening up global-North-led research on sustainable-energy transitions in global South contexts to examine the underlying assumptions, blind spots and limitations, in an attempt to identify alternative ways of doing transitions research that are more responsive to the priorities of the transitioning communities. As an example of one of these alternative ways of doing transitions research, the SONG project aims to generate sustainable-energy solutions and transformative change in two Kenyan communities through the co-design of socio-technical systems on the basis of bottom-up iterative community engagement and the integration of social-science and engineering research. The SONG project's ambition is novel in that it has taken the theoretical concept of niche experimentation and applied it in practice to establish community innovation niches to facilitate energy transitions.

A review of the socio-technical transitions literature, predominantly focused on the experiences of industrialised countries in the global North, identified a number of conceptual limitations that can be broadly construed as a neglect of issues of distributive and procedural justice that limit the ability of transitions research and its implementation to provide just and equitable solutions for transitioning communities in the global South. Addressing two prominent justice deficits among these – a lack of consideration of the politics of framing and of the socio-spatial relations and dynamics – was considered fundamental to achieving socially just energy transitions in the specific contexts of the community niches in Kenya. The co-design of the SONG project can thus be seen as a response to the limitations and justice deficits of the dominant transitions frameworks.

Attention to the principles of distributive justice related to socio-spatial relations and dynamics in the two community niches was important to ensure that inequalities driven by age, gender, ethnicity and income were alleviated and not exacerbated by the SONG project. Issues of intra- and intergenerational justice, while absent in transition frameworks, were nonetheless important in the SONG project for drawing attention to the vital roles of gender and age in co-designing socio-technical systems that are responsive to different intra- and intergenerational priorities and aspirations, while promoting the equitable distribution of potential benefits and risks across space and time.

The lens of procedural justice spotlighted the heightened politics of transitions in the global South, and reminded the SONG project team of the need for opportunities that are both inclusive and interactive so that community members can participate in the script of their own community energy services, be co-designers in the shaping and delivery of those services, and thus be the co-producers of their own energy transitions. The case of Echareria has shown the need to take into account the political issue of community non-cohesion, to prevent the reproduction or exacerbation of any tensions between different social groups and to enhance the community's capacity to decide on a common energy future. If energy transitions for development are to be truly transformative and just, then they must incorporate opportunities for the community to develop local governance processes and capacity building that are respectful of existing hierarchical governance structures, but extend participation beyond them, to ensure the effective and inclusive management of transitions to sustainable energy futures.

In conclusion, this chapter began with the premise that making transitions research 'open' from the bottom up presupposes the embedding of democratic values such as inclusiveness and social justice. In practice, however, translating Northern concepts and assumptions into global-South contexts highlighted the challenges of social justice and spatial positioning that underlie attempts to co-design transitions research that spans distinct geographical and socio-cultural boundaries.

References

Berkhout, F., Verbong, G., Wieczorek, A., Raven, R., Lebel, L., and Bai, X. (2010). Sustainability experiments in Asia: Innovations shaping alternative development pathways? *Environmental Science and Policy*, 13(4), 261–271.

Byrne, R., Ockwell, D., Urama, K., Ozor, N., Kirumba, E., Ely, A., Becker, S., et al. (2014). *Sustainable Energy for Whom? Governing Pro-poor, Low Carbon Pathways to Development: Lessons from Solar PV in Kenya.* STEPS Working Paper 61. Brighton: STEPS Centre.

Byrne, R. P. (2011). Learning Drivers: Rural Electrification Regime Building in Kenya and Tanzania. PhD thesis, University of Sussex.

Coenen, L., Benneworth, P., and Truffer, B. (2012). Toward a spatial perspective on sustainability transitions. *Research Policy*, 41, 968–979.

Eames, M., and Hunt, M. (2013). Energy justice in sustainability transitions research. In K. Bickerstaff, G. Walker and H. Bulkeley (eds), *Energy Justice in a Changing Climate: Social Equity and Low-Carbon Energy* (pp. 46–60). London: Zed Books.

Fuller, S., and Bulkeley, H. (2013). Energy justice and the low-carbon transition: Assessing low-carbon community programmes in the UK. In K. Bickerstaff, G. Walker and H. Bulkeley (eds), *Energy Justice in a Changing Climate: Social Equity and Low-Carbon Energy* (pp. 61–78), London: Zed Books.

Geels, F. W. (2005). *Technological Transitions and System Innovations: A Co-evolutionary and Socio-Technical Analysis.* Cheltenham: Edward Elgar.

Geels, F. W. (2010). Ontologies, socio-technical transitions (to sustainability), and the multi-level perspective. *Research Policy*, 39(4), 495–510.

Geels, F. W. (2011). The multi-level perspective on sustainability transitions: Responses to seven criticisms. *Environmental Innovation and Societal Transitions*, 1(1), 24–40.

Geels, F. W., and Raven, R. P. J. M. (2006). Non-linearity and expectations in niche-development trajectories: Ups and downs in Dutch biogas development (1973–2003). *Technology Analysis and Strategic Management*, 18, 375–392.

Geels, F. W., and Schot, J. (2007). Typology of sociotechnical transition pathways. *Research Policy*, 36(3), 399–417.

Goldthau, A., and Sovacool, B. K. (2012). The uniqueness of the energy security, justice, and governance problem. *Energy Policy*, 41, 232–240.

Lake, R. W. (1996). Volunteers, nimbys, and environmental justice: Dilemmas of democratic practice. *Antipode*, 28(2), 160–174.

Lawhon, M., and Murphy, J. T. (2011). Socio-technical regimes and sustainability transitions: Insights from political ecology. *Progress in Human Geography*, 36(3), 354–378.

Lehtonen, M., and Kern, F. (2009). Deliberative socio-technical transitions. In I. Scrase and G. MacKerron (eds), *Energy for the Future: A New Agenda* (pp. 103–122), Basingstoke: Palgrave Macmillan.

Meadowcroft, J. (2009). What about the politics? Sustainable development, transition management, and long term energy transitions. *Policy Sciences*, 42(4), 323–340.

Newell, P., and Mulvaney, D. (2013). The political economy of the 'just transition'. *Geographical Journal*, 179(2), 132–140.

Ockwell, D., and Byrne, R. (2015). Improving technology transfer through national systems of innovation: Climate relevant innovation-system builders (CRIBs). *Climate Policy*, 16(7), 836–854 [epub ahead of print], doi: 10.1080/14693062.2015.1052958.

Okereke, C. (2010). Climate justice and the international regime. *WIREs Climate Change*, 1, 462–474.

Pulido, L. (1996). *Environmentalism and Social Justice: Two Chicano Struggles in the Southwest*. Tucson, AZ: University of Arizona Press.

Raman, S., and Mohr, A. (2013). Biofuels and the role of space in sustainable innovation journeys. *Journal of Cleaner Production*, 65, 224–233.

Raven, R., Schot, J., and Berkhout, F. (2012). Space and scale in socio-technical transitions. *Environmental Innovation and Societal Transitions*, 4, 63–67.

Scharpf, F. (1999). *Governing in Europe: Effective and Democratic?* Oxford: Oxford University Press.

Schlosberg, D. (2007). *Defining Environmental Justice: Theories, Movements, and Nature*. Oxford: Oxford University Press.

Smith, A., Kern, F., Raven, R., and Verhees, B. (2014). Spaces for sustainable innovation: Solar photovoltaic electricity in the UK. *Technological Forecasting and Social Change*, 81, 115–130.

Smith, A., and Stirling, A. (2007). Moving outside or inside? Objectification and reflexivity in the governance of socio-technical systems. *Journal of Environmental Policy and Planning*, 9(3–4), 351–373.

Smith, A., Voß, J.-P., and Grin, J. (2010). Innovation studies and sustainability transitions: The allure of the multi-level perspective and its challenges. *Research Policy*, 39(4), 435–448.

Stirling, A. (2014). *Emancipating Transformations: From Controlling 'the Transition' to Culturing Plural Radical Progress*. STEPS Working Paper 64, Brighton: STEPS Centre.

van der Horst, D., and Vermeylen, S. (2011). Spatial scale and social impacts of biofuel production. *Biomass and Bioenergy*, 35(6), 2435–2443.

Wang, T., and Watson, J. (2010). Scenario analysis of China's emissions pathways in the 21st century for low carbon transition. *Energy Policy*, 38(7), 3537–3546.

8

Monstrous regiment versus Monsters Inc.: competing imaginaries of science and social order in responsible (research and) innovation

Stevienna de Saille, Paul Martin

> All monsters are undead. Maybe they keep coming back because they still have something to say or show us about our world and ourselves. Maybe that is the scariest part. (Beal, 2014: 10)

As new technological domains emerge, so too do promises and warnings about the future they will bring. However, as technology has grown ever more complex, predicting either benefits or risks has become increasingly difficult, particularly where a high level of control over natural processes is concerned. This can lead to uneasiness on the part of both the public and policymakers that scientific information alone cannot counter (Douglas, 2015). Moreover, in a time of rapidly increasing political and economic uncertainty, the difficulty of separating science from politics, and business interests from both, only exacerbates the public's concern that necessary precautions may be discarded in the rush to open up new markets and find new sources of economic growth.

Powerful corporations do indeed succeed in shaping regulations to their needs, at times with great cost to public health or the environment. But potentially useful technologies can also be blocked by a public that does not trust those who create and regulate technology to hear its concerns. For some technologies these tensions have resulted in an intractable, decades-long debate in which each side imagines the other as monstrous and rampaging out of control. In this chapter we want to open up these ideas, to examine how the concept of 'monsters' functions in entrenched controversy, and ask whether this can be addressed through evolving frameworks for responsible (research and) innovation (RI).[1]

Since the 1980s it has been suggested that greater understanding of science (and, later, greater engagement with science) would help provide the context necessary for the public to feel confident of the benefits of a new technology, as well as mitigating fear of its risks. Historically, however, these imperatives to engage the public as a means of gaining legitimacy for new technological fields have been accompanied by a growing fear of those who wish to engage with science on their own terms (Hess, 2014; Marris, 2014; Welsh and Wynne, 2013), in particular through direct protest. While considering RI as a more responsible way of 'embedding' innovation in society (von Schomberg, 2014: 39), it is therefore also necessary to consider the ways in which it represents a socio-technical imaginary in which progress is achieved through enlightened governance of the entire research and innovation system, in order to keep under control both potential technological monsters and a potentially unruly public response.

According to anthropologist Martijntje Smits (2006), monsters embody inseparable but directly opposing characteristics that make them both horrifying and fascinating at the same time. Because the simultaneous fear and attraction they evoke can never be reconciled, monsters are more than mere creatures of the imagination; they are deeply rooted in the binaries of modernism – body versus mind, nature versus technology, superstition versus science. When we speak of monsters in the socio-technical imaginary in this sense, we are referring not to the 'imaginary' as non-existent, but to a concrete, collectively embraced, politically actionable future in which technology is expected to bring about certain positive, culturally intelligible improvements to the human condition (Jasanoff, 2015; Jasanoff and Kim, 2009). These are generally (although not only) framed as large-scale scientific projects that embody national economic aspirations – for example, building a civil nuclear industry or setting out a strategy for world leadership in a new scientific field. As socio-technical imaginaries are also rooted in the modernist project of social progress

---

1 We use RI here to refer generically to discussions around responsible innovation, and RRI to refer specifically to its recent formulation by the European Commission as part of Horizon 2020, its funding programme for research and innovation.

through science, arguments for and against these projects also tend to be framed through binary opposition, so that science, rationality and control wind up juxtaposed against culture, emotion and messy nature. Monsters, therefore, can be thought of as signifiers of disorientation in imaginaries of progress, markers for that which cannot easily be assigned to one side of the binary or the other, perhaps cannot even be properly categorised at all because they too are unknown, like the warnings placed over the uncharted portion of an incomplete map. To illustrate these points more clearly in the discussion that follows, we will draw upon both the Frankenstein story, as one of the original monsters in the socio-technical imaginary of progress through science, and more recent metaphors from popular culture, to discuss genetically modified organisms (GMOs), one of the most intractable technological controversies of our time.

While those involved in creating GMOs, particularly multinational corporations such as Monsanto, have been characterised as a kind of 'Monsters Inc.', likewise the biotechnology sector has sometimes appeared to view the public as a kind of monstrous regiment, an army of dissent intent on thwarting the aspirations of a field that seeks only to improve upon nature to feed the world (Riley-Smith, 2014). We draw the metaphor from the highly successful animated Pixar film *Monsters, Inc.*, released by Disney in 2001 (Docter et al., 2001). The film turns on the idea that the monsters hiding in our childhood closets have a reason for terrifying us every night: power for their city, Monstropolis, is derived from human children's screams. Unfortunately, modern children are not as fearful as they once were, and a scream shortage is now threatening the economic security of Monsters Inc., the factory which produces the 'clean energy' that powers the city. The conceit of the story is that the monsters are really just like us – worried about jobs, family and social status – and it is human children who are monstrous, radioactively toxic creatures who must be contained behind their closet doors lest they contaminate the normal world. When a human child (Boo) sneaks into the factory on the back of her 'scarer' (Sulley), her discovery throws the whole city into a panic. In solving the problem of getting Boo safely back to her world, Sulley has to reconcile his fear of human children with the realisation that, as a top scarer, he is the monster in Boo's eyes. This reconciliation ultimately leads the monsters to discover that children's laughter is a more powerful energy source than screams.

Ultimately, as a children's film, *Monsters, Inc.*, validates cooperation and understanding as leading to better ways of producing what is necessary for society to prosper. However, as Tranter and Sharpe (2008: 305) point out, it is also a parable about fear, both of running out of the energy upon which society depends, and – in the character of the CEO Waternoose, who is plotting to kidnap human children and torture them to produce more efficient screams – of how far some corporations may go to secure a profit. The film's references to clean energy and a running joke about the vicious scrubbing to which human-contaminated monsters are subjected draw heavily on cultural knowledge about the civil nuclear industry, where clean energy cannot be separated from the deadly toxicity of the materials used and the waste produced, even were accidents never to occur. Boo, therefore, functions as both a caution against excessive fear (she is not actually toxic), but also as a reminder that technology can have undesirable social costs.

Unlike Monsters Inc., the term 'monstrous regiment' is several centuries old. It originates with a sixteenth-century Protestant reformer who coined the phrase to argue against the passage of the Scottish Crown to a woman, the Catholic Mary Queen of Scots (Knox, 2011 [1558]). 'Regiment' in this time meant rule, not army. The monstrosity in this case was allowing a woman to assume a role that God had reserved for men. Terry Pratchett (2003) later used the term more literally in a satiric novel about a Balkans-like country engaged in an endless war no-one seemed to understand, fought by an army of women secretly disguised as men (the actual men having all been captured or run away). We use the term both to characterise the illusion of an army which does not actually exist and, 'monstrous regiment' being inherently a gendered term, the practice of ascribing stereotypically 'feminine' negative traits – such as being irrational, emotional and illogical – to those who attempt to oppose regimes of power such as science, technology, politics and the market, which are still very much the province of the male.

Beginning with an exploration of metaphors as carriers of meaning, we will then use the metaphors of Monsters Inc. and the monstrous regiment to consider how certain actors and technologies become constructed as monstrous, and how this contributes to the continued intractability of some technological debates. We end by asking what 'monster theory' (Smits, 2006) can offer towards a better understanding of competing social imaginaries in RI.

Monsters in the imaginary (rather than imaginary monsters)

Metaphors function as condensed, culturally intelligible stories, or 'scripts' (Shank, 1990, quoted in Turney, 2000: 6) that guide meaning making towards a prespecified conclusion. Perhaps their most visible use in controversy is through symbolic action by social movements, in order to illustrate an issue that has been left unspoken (Melucci, 1985). For example, anti-genetic modification (anti-GM) protesters often wear hazmat suits to illustrate the toxicity of pesticides used with GM crops. In particular, metaphors are able to carry meaning from one context to another (Hellsten, 2006), transferring or extending the original script to make a clear statement about how to read something new.

Inasmuch as monsters can be defined as an irreconcilable hybrid of two opposing cultural categories simultaneously eliciting both fear and wonder (Smits, 2006), they do not spring purely from wild imagination. Rather, as Smits argues, monsters require grounding in a culturally specific context to give them concrete existence. The script for Frankenstein is generally interpreted as a reflection of the uneasiness with which Mary Shelley and her circle of friends viewed the pace of change from agrarian society to industrial modernity, and it is still considered a cautionary tale about the capacity of technology to do ill as well as good (Botting, 2003; Latour, 2012; Turney, 2000). As Turney explains:

> It is frightening because it depicts a human enterprise which is out of control, and which turns on its creator ... [but] the myth is never a straightforward anti-science story ... the *Frankenstein* script, in its most salient forms, incorporates an ambivalence about science, method and motive, which is never resolved. (Turney, 2000: 5)

This ambivalence is also apparent in the public's tendency to amalgamate the scientist, Victor Frankenstein, with the flat-headed monster of

Hollywood films. As Smits (2006) argues, the creature's monstrosity is not merely the result of being a crime against nature, as other commentators suggest, but rather the internal chaos caused by the contradictions of being simultaneously alive and dead, benign and lethal, natural and technological. The maker of the creature is equally reviled and persecuted as a man seeking the power of granting life, which is supposed to belong only to God, but whose horrified abandonment of his creation sets a series of violent events in motion. In other words, both the creature and the scientist embody key binaries of modern progress which can never be reconciled by calling upon either values or science (Smits, 2006: 499). For Shelley, the context was the development of electricity and the industrial revolution. Cultural theorists have similarly argued that the surge in the prevalence of monstrous characters and topics in popular media post-9/11 signifies that the monster figure may have now 'transcended its status as metaphor' to become 'a necessary condition of our existence in the 21st century' as both technology and social relations become ever more complex (Levina and Bui, 2013). Zombies, for example, which were once slow undead shufflers, are now victims of a virus capable of super-fast motion and driven by super-rage,[2] perhaps reflecting a shift in the locus of public fear from the supernatural to the technological, and from mindless violence to directed fury. As in Shelley's work, therefore, monsters continue to reflect unease not only with technology per se but with the increasing pace and direction of change. Similarly, historical uprisings against the introduction of new technologies, in particular the Luddites smashing the first mechanised looms as a threat to their jobs and way of life as traditional weavers, are still featured in the techno-scientific imaginaries of modernity as ignorant, superstitious mobs standing in the way of inevitable progress (see, for example, Wood, 1999). They are the pitchfork-wielding villagers of the Hollywood version of the Frankenstein story, whose rejection sets the creature on its rampage, so that the cultural resonance of the monstrous regiment is always anti-progress and anti-science.

These two inherently irreconcilable scripts, that of technology as simultaneously the saviour of humanity and a kind of monster factory,

---

2 As depicted in a new generation of films such as *28 Days Later* (20th Century Fox, 2002) and *World War Z* (Paramount Pictures, 2013).

and the public as simultaneously the eager beneficiary and the irrational (yet organised) saboteur of technological progress (Marris, 2014; Nowotny, 2014; Welsh and Wynne, 2013) are extremely powerful in the debates on GMOs. But is it possible to use the values that RI supposedly embodies to directly grapple with – rather than seek to bury – the contradictions that are inherent in GM itself?

Monsters and the imaginary of RI

The socio-technical imaginary of RI – in which research and innovation, guided by the continual involvement of a technologically literate public, result in goods and services that answer society's needs – has been the most recent attempt to answer unease about the risk and pace of change. While there is no agreed definition of RI, in general it is understood as a means of involving the public in an iterative and mutually reflexive process of shaping innovation, from upstream research, all the way downstream to its eventual introduction to the market and beyond (Owen et al., 2013; von Schomberg, 2013). However, it is also in the interests of multinational corporations and growth-eager governments to shape public engagement in ways that tend to dismiss, defuse and disinvite engagement from the dissenting public (de Saille, 2015) before opposition can grow strong enough to interfere with the commercialisation process; in effect producing a good public for the product rather than a product for the public good (Thorpe and Gregory, 2010). This has also been an irreconcilable contradiction in the development of RI.

Lurking deep within the imaginary of RI is the monster of recombinant DNA (rDNA), born in the early 1970s to both fanfare and fear, not least amongst the scientists, who were worried that public reaction to it might curtail their research. If, in this story, rDNA was an infant monster, the lullaby that sang it to sleep was the 1975 Asilomar Conference on Recombinant DNA, at which genetic scientists, accompanied by lawyers, journalists and government officials, agreed on a set of restrictions within which they could continue their work (Berg, 2008). Acknowledging that there was particular concern about rDNA's potential siblings – viral cloning and bacterial pathogens – scientists set about reassuring the public that the process of recombining DNA to create organisms not found in nature would only ever be used for good, not harm, and that these could be contained within the lab until mature enough to be safely released.

By the early 1980s the field of agricultural GM was slowly being made 'business-friendly' through deregulation (Stavrianakis, 2012: 159), and in 1994, with the introduction of the Flavr Savr tomato to the US market, rDNA's children finally began to leave the lab and seek their fortune in the world. The Flavr Savr was engineered by Calgene to turn off the genes responsible for softening. Ordinary tomatoes must be picked and shipped while they are hard and still green, and are then chemically ripened with ethylene gas, but these could remain firm while they matured on the vine. This improved flavour so much that despite being almost twice the price of ordinary tomatoes they met with good public demand.[3] Ultimately though, Calgene was staffed by scientists, not farmers, who did not know how to get the tomatoes from the fields to the market in numbers large enough to make the business profitable.[4] In 1997 the company and its patents were acquired by Monsanto, which promptly discontinued production of the Flavr Savr, as it had its own tomato under development. A paste made from GM tomatoes was launched on the UK market in 1996 but withdrawn in 1998 when sales fell sharply after a media broadcast claimed that while the modified gene itself was not a risk, adverse health effects could still ensue from the GM process (Bruening and Lyons, 2000).

According to most of the tellers of this tale, this is when the European public, having until then been largely content with Asilomar's assurance that science could be trusted to raise the monster with care, suddenly panicked and grabbed their torches and pitchforks with the intention of driving GMOs from their shores. Organic farmers' associations, in particular, were increasingly concerned about pollen contamination from adjacent GM fields, as this could cause them to lose their certification and thus their entire business. In 1998 a leading organic farmer in the UK sued for a judicial review of the legality of a maize test site adjacent to his own organic maize. He eventually won the case on

---

3 One Calgene scientist interviewed twenty years later remembers they were so popular in her hometown of Davis, California, that greengrocers had to ration them (for a video report, see Winerip, 2013).
4 Again, for a video report see Winerip (2013).

appeal, but when the GM crop was not ordered to be removed, his supporters invaded the field and destroyed the plants before their pollen could be released (Reed, 2002). In 2000 a trial jury accepted that the destruction was done 'in order to prevent greater harm', setting a precedent which has meant that for most similar instances since then, crop destruction has gone unpunished (Tait, 2009). Activists have subsequently destroyed more than eighty fields and test sites in Europe alone (Kuntz, 2012), have staged numerous protests across the world, and have made both carefully documented and wildly inaccurate claims about health and environmental risks, soil depletion, threats to bee and butterfly populations, predatory corporate practices, and social upheaval caused by adoption of GMOs.

The monstrous regiment, therefore, has certainly engaged in behaviour that is reminiscent of pitchforks and torches. In a more recent example, Ecover, a company known for its natural products, unexpectedly ran into strong resistance when it announced that for reasons of sustainability it planned to replace the palm oil in its laundry detergent with an oil derived from genetically modified algae (Strom, 2014). Although initially stating that the company that supplied the algal oil was using synthetic biology to produce it, when protest ensued, some representatives subsequently tried to argue that synthetic biology meant creating a new organism from scratch, whereas they were only harnessing the algae's 'natural' fermentation process (Domen and Develter, 2014). Ecover, in fact, had been using enzymes created from GM bacteria for some time (see Asveld and Stemerding, 2016, for a full account), but it is possible that consumers were not aware of this before. Neither argument was likely to defuse resistance to the product among Ecover's deep green customers, who could not reconcile the idea of scientifically altering an organism to produce something it did not normally produce with the term 'natural' (Thomas, 2014). A boycott was called, after which Ecover withdrew the product and announced a period of public consultation. As of the end of 2016, any further plans to use algal oil appear to have been shelved.

In terms of the socio-technical imaginary of public engagement in innovation through RI, it could be said that this is how RI could and should function, in that the incident created an open public debate and the company responded to concern about a pathway of innovation and changed its trajectory accordingly. However, few on either side would claim to be satisfied with the outcome, particularly as the laudably responsible goal of replacing palm oil with a more sustainable alternative remains unrealised. Ecover's experience suggests that the difficulties of innovating responsibly are far more complex than simply 'engaging' the public, and that even a company with an ethical reputation could suddenly find itself seen as Monsters Inc. if it tried to smooth away the irreconcilable tensions between the natural and the technical that are inherent in genetic engineering.

However, focusing on the irreconcilable tensions of the technology itself should not obscure acknowledgement of fear and fascination also arising from the context within which a new technology will be received. Monsanto, for example, has been extremely successful at commercialising GM crops as the solution to feeding a growing world population. But it has also been called 'Monster Monsanto' because of its historic production of Agent Orange, a defoliant used to destroy North Vietnamese agriculture during the Vietnam War, which has had long-term health effects on both US soldiers (and their descendants) and the Vietnamese population (see, for example, Ragonesi, 2004). Although Monsanto stopped producing Agent Orange for the military in 1969, its website states that it is still 'responsible' for the compound (Monsanto, 2017), whose broad herbicidal function is similar to its signature product, Roundup, which kills all plants except the seeds engineered to withstand it. Roundup's active ingredient, glyphosate, has been the subject of intense scientific debate about long-term toxicity,[5] while Monsanto was also a major producer of polychlorinated biphenyls (PCBs) and dioxin, and has been sued repeatedly (and occasionally even successfully[6]) for contamination of dozens of sites in the USA alone. Therefore, beyond the ethical questions which have often surrounded Monsanto's particular model of generating profit

---

5 The World Health Organization's International Agency for Research on Cancer (IARC) has issued a report reclassifying glyphosate as 'probably' carcinogenic (IARC, 2016), causing some member states to refuse to renew its licence for sale in the European Union. The European Food Standards Agency (EFSA) subsequently issued its own report finding it not carcinogenic if used as recommended (EFSA, 2015). As of this writing, the licence has been temporarily renewed while a third agency reports (European Commission, 2016).
6 As one high-profile example, a long-standing lawsuit over clean-up in Anniston, Alabama, was finally settled for $700 million (Associated Press, 2003).

through the stringent enforcement of patents (Barlett and Steele, 2008) or by successfully lobbying the US Congress for laws that protect it from being sued in the public interest (Godoy, 2013), the company's reputation as a food producer continues to be inseparable from its historic reputation and present market leadership as a producer of poison. Moreover, from the point of view of the public, the way in which Monsanto has pursued the commercialisation of GM can make all industrial actors in the field look like Monsters Inc., as well as making any application of GM, however useful, seem monstrous – as with the term Frankenstein, the creator has been conflated with its creation over time.

At the same time, the failure to commercialise Golden Rice, which has been hailed as a 'humanitarian' GM crop aimed at delivering vitamin A to areas which are traditionally malnourished, has been blamed on the 'wicked' monstrous regiment of anti-GMO activists destroying test fields (Riley-Smith, 2014), with some GM proponents even circulating images of starving children, captioned 'Thank you, Greenpeace, for saving us from Golden Rice' (Everding, 2016). Embarking on a campaign of public 'education' about GM in the 1980s, the chairman of Monsanto similarly said that:

> the only thing that will stand in the way of our achieving the full potential of our next golden era is that we will be thwarted by a public that doesn't understand science or technology and that doesn't trust us to use science wisely and with appropriate regard for the concerns of the public we serve. (Richardson, 1985, quoted in Kleinman and Kloppenburg, 1991)

The above statement resonates with the argument, common in the 1980s and 1990s, that the public was resistant to new technology because there was a 'deficit' of scientific understanding. Trust in the company and understanding of the science are conflated, so that better science communication is assumed to produce trust in the company's agenda, and when this does not happen, it is the public's ignorance that continues to be blamed. Yet some scientists working on the rice have insisted that the problems are scientific, not political, and in the meantime the Philippines has reduced vitamin A deficiency by methods which do not include GMOs (Everding, 2016).

These examples suggest competing imaginaries of social order, one in which progress through technology is inevitable and desirable, no matter the cost, and the other in which technology is not always benign, nor even always necessary for social progress. As technology becomes ever more complex, finding means through which we might be able to explore both the fascination and the fear becomes ever more crucial, not least to the imaginary of RI.

Monster theory, or bringing the monsters into the open

To return to our metaphors, from the point of view of Monsters Inc., the human world (the public) is both potentially toxic and necessary to produce the energy (of market acceptance) upon which successful innovation depends. In this imaginary the public does not understand either science or the importance of Monsters Inc. as a driver of the economy, and so is standing in the way of progress and growth. This imaginary, however, can also perceive 'the public' as more singular and unified than it actually is. In the case of rDNA's great-grandchild, synthetic biology, Marris (2014) suggests there is now a kind of synbiophobia-phobia, in which an exaggerated fear of a potential public backlash informs the drive towards RI. Marris argues that the continued assumption that resistance is a product of ignorance has meant that, in fact, the 'inclusive engagement' elements of RI have remained mired in precisely the kind of top-down 'we talk, you listen' form of scientific communication that RI was meant to overcome. Science-studies scholars have long criticised such endeavours, suggesting instead that the public is not a single entity, but is made up of many publics with many different sets of values that come into play. These publics need to have opportunities for meaningful engagement with scientific topics, which includes the right to introduce questions about the social impact of technological change, rather than simply being furnished with more scientific information.

In *Options for Strengthening Responsible Research and Innovation*, the European Commission (2013: 14) agrees that the €300 million spent on safety research has not been adequate to resolve society's concerns, and laments the lack of a European market for GMOs. In this and other key policy documents, RRI is constructed as a way to resolve these concerns at an early stage, so as to accelerate the pace of innovation without incurring the kind of intractable distrust and resistance which has been the case with GMOs. However, key assumptions linking innovation to growth, and both to social progress, remain unquestioned, as does the linkage between upstream information and downstream trust. As a tool for shaping avenues to facilitate meaningful engagement, it is possible that RI's drive towards openness is more likely to produce an artificial closure at the earliest stages of the development process, thus re-entrenching old positions, rather than ameliorating them, as it is how technology acts in the world – what Frankenstein's creature does after exiting the lab, more than the fact of his existence – that matters most in terms of generating trust.

Braidotti (1996: 135) suggests that 'to signify potentially contradictory meanings is precisely what the monster is supposed to do'. In other words, it is perhaps not the intractability of polarised positions that is the problem, but the reluctance to honestly acknowledge that the contradictions cannot be prised apart. Scarers on the factory floor do not see themselves as monsters, but rather as the responsible producers of a necessary commodity. Likewise, human children are still needed for the energy they provide, but this can be produced in a better way once it becomes apparent that they are not as monstrous as the residents of Monstropolis fear.

In an earlier definition of the term, Law (1991) even goes so far as to suggest that 'socio-technical' is a monster in and of itself, as the social and the technical are generally seen as occupying mutually exclusive disciplinary realms. This suggests that 'society' exists in permanent opposition to science, and indeed much of the polarisation in the GM debate is due to a framing that relegates social and ethical concerns to the irrational, set in opposition to rational concerns that are usually framed as economic.[7] In this way, by not continuing to acknowledge both the fear and the fascination as legitimate responses, corporations engaged in biotechnology are perceived as untrustworthy and not interested in what the consumer might really want to purchase (such as better-tasting tomatoes), but rather only in advancing the interests of industrial agriculture and their own profits (Holland, 2016).

Latour (2012) argues that the dominion and mastery that signify progress do not bring freedom from nature, but rather an ever-deepening attachment to that which is being dominated and mastered.

---

7 See, for example, arguments that the 'green blob' of environmentalists are acting against national economic interests (Riley-Smith, 2014).

He argues that 'we use the monster as an all-purpose modifier to denote technological crimes against nature' because, like Victor Frankenstein, we are unable to either love our technological creations or accept that it is our own actions that turn good creations bad. Plastics, for example, were initially hailed as a wonder (Smits, 2006). They became monstrous only when it became apparent that they continued to retain their cultural shaping (such as bags and bottles) after they became waste, causing great damage out of context by, for example, clogging up waterways and strangling birds. In other words, RI suggests that we must not, like Victor Frankenstein, run away from the technological monsters we have created; we must love and socialise and remain always responsible to our monsters so we may live with them in peace. But loving and taming our monsters may first require that we are able to acknowledge that 'monsters' is exactly what they are.

Conclusions

RI claims that greater public involvement at the research stage will shape innovation towards social benefit, but it has less capacity to control what happens when the monsters leave the lab and begin to interact with the world. As we cannot separate trust in science from trust in the adequate regulation of its marketable technologies, attempts to provide more scientific information about monsters which have not yet left the lab often produces no difference in acceptability and may, in fact, provoke more distrust as the inbuilt tensions between claimed social benefit and real-world actions, and between corporate risk management and corporate profit, become clear.

For Smits (2006), taming the monster – what she calls the cultural domestication of a new technology – is a process of mutual shaping, so that the technology opens a new space in the existing social order even as the social order tames the technology to fit it. Indeed, not all monsters are necessarily the stuff of nightmares – mermaids and satyrs, for example, while embodying contradictions, are not nearly as unsettling as Frankenstein's creature or the fast zombie. Considering controversial technologies through Smits's monster theory may reveal some of the simultaneous contradictions inherent not only in the innovation itself, but in the entire shaping of the innovation system towards the production of economic growth even at great social cost.

Whether the monster enables desirable or undesirable outcomes – whether it becomes Sully or Waternoose – will require something more than controlled forms of public engagement to provide a context in which it can be adequately socialised. RI should not be aimed at resolving differences, but rather seeking new forms of openness that can accept contradictions and explore the social and the technical as intertwined. Instead of leading to intractable polarisation, bringing the monsters out of the shadows through RI may help point to more useful ways of dealing directly with their potential to turn on their creators, thus attempting to productively address larger concerns about the overall pace and trajectory of technological change. We have suggested that the socio-technical imaginary of RI still contains a fear of both Monsters Inc. and the pitchfork-wielding monstrous regiment, the truth and exaggeration of which must be brought into the open, as part of the landscape in which innovation occurs. Taming, rather than hiding, the monster could be key to realising RI's goal of creating a new socio-technical imaginary of governance which enables truly socially beneficial research and innovation to emerge.

References

Associated Press (2003). $700 million settlement in Alabama PCB lawsuit. *New York Times*, 21 August. Retrieved 8 July 2016 from: www.nytimes.com/2003/08/21/business/700-million-settlement-in-alabama-pcb-lawsuit.html.

Asveld, L., and Stemerding, D. (2016). *Algal Oil On Trial: Conflicting Views of Technology and Nature*. The Hague: Rathenau Instituut.

Barlett, D. L., and Steele, J. B. (2008). Monsanto's harvest of fear. *Vanity Fair*, 2 April. Retrieved 29 February 2016 from: www.vanityfair.com/news/2008/05/monsanto200805.

Beal, T. K. (2014). *Religion and Its Monsters*. New York: Routledge.

Berg, P. (2008). Meetings that changed the world: Asilomar 1975 – DNA modification secured. *Nature*, 455(7211), 290–291.

Botting, F. (2003). Metaphors and monsters. *Journal for Cultural Research*, 7(4), 339–365.

Braidotti, R. (1996). Signs of wonder and traces of doubt: On teratology and embodied differences. In N. Lykke and R. Braidotti (eds), *Between Monsters, Goddesses and Cyborgs: Feminist Confrontations with Science, Medicine and Cyberspace* (pp. 135–152). London: Zed.

Bruening, G., and Lyons, J. (2000). The case of the FLAVR SAVR tomato. *California Agriculture*, 54(4), 6–7.

de Saille, S. (2015). Dis-inviting the unruly public. *Science as Culture*, 24(1), 99–107.

Docter, P., Silverman, D., and Unkrich, L. (2001). *Monsters, Inc.* Burbank/Emeryville, CA: Walt Disney Pictures/Pixar.

Domen, T., and Develter, D. (2014). Ecover is as green as ever! *Ecologist* [blog]. Retrieved 8 July 2016 from: www.theecologist.org/blogs_and_comments/Blogs/2450666/ecover_is_as_green_as_ever.html.

Douglas, H. (2015). Politics and science: Untangling values, ideologies, and reasons. *Annals of the American Academy of Political and Social Science*, 658(1), 296–306.

EFSA (2015). Conclusion on the peer review of the pesticide risk assessment of the active substance glyphosate. *EFSA Journal*, 13(11), 4302.

European Commission (2013). *Options for Strengthening Responsible Research and Innovation*. Brussels: European Commission.

European Commission (2016). FAQs: Glyphosate [update 29 June]. *European Commission Press Release Database*. Retrieved 8 July 2016 from: http://europa.eu/rapid/press-release_MEMO-16-2012_en.htm.

Everding, G. (2016). Genetically modified Golden Rice falls short on lifesaving promises: GMO activists not to blame for scientific challenges slowing introduction, study finds. *Source*, 2 June, Washington University in St. Louis. Retrieved 31 June 2016 from: https://source.wustl.edu/2016/06/genetically-modified-golden-rice-falls-short-lifesaving-promises.

Godoy, M. (2013). Did Congress just give GMOs a free pass in the courts? *National Public Radio*, 21 March, 5.59 PM ET. Retrieved 17 August 2016 from: www.npr.org/sections/thesalt/2013/03/21/174973235/did-congress-just-give-gmos-a-free-pass-in-the-courts.

Hellsten, I. (2006). Focus on metaphors: The case of 'Frankenfood' on the web. *Journal of Computer-Mediated Communication*, 8(4), 0, doi:10.1111/j.1083-6101.2003.tb00218.x (originally published 2003).

Hess, D. J. (2014). Publics as threats? Integrating science and technology studies and social movement studies. *Science as Culture*, 24(1), 69–82.

Holland, N. (2016). Biotech lobby's push for new GMOs to escape regulation. *Corporate Europe Observatory*. Retrieved 8 July 2016 from: http://corporateeurope.org/food-and-agriculture/2016/02/biotech-lobby-push-new-gmos-escape-regulation.

IARC (2016). *Monographs on the Evaluation of Carcinogenic Risks to Humans*. Volume 112: *Glyphosate* [updated 11 August 2016]. Lyon: IARC. Retrieved 4 December 2016 from: http://monographs.iarc.fr/ENG/Monographs/vol112/index.php.

Jasanoff, S. (2015). Future imperfect: Science, technology, and the imaginations of modernity. In S. Jasanoff and S.-H. Kim (eds), *Dreamscapes of Modernity:*

Sociotechnical Imaginaries and the Fabrication of Power (pp. 1–33). Chicago: University of Chicago Press.

Jasanoff, S., and Kim, S.-H. (2009). Containing the atom: Sociotechnical imaginaries and nuclear power in the United States and South Korea. *Minerva*, 47(2), 119–146.

Kleinman, D. L., and Kloppenburg, J. J. (1991). Aiming for the discursive high ground: Monsanto and the biotechnology controversy. *Sociological Forum*, 6(3), 427–447.

Knox, J. (2011 [1558]). *The First Blast of the Trumpet against the Monstrous Regiment of Women* (html version, taken from 1878 edition, ed. by E. Arber; originally published 1558). Retrieved 2 March 2016 from: www.gutenberg.org/files/9660/9660-h/9660-h.htm.

Kuntz, M. (2012). Destruction of public and governmental experiments of GMO in Europe. *GM Crops and Food*, 3(4), 258–264.

Latour, B. (2012). Love your monsters: Why we must care for our technologies as we do our children. *Breakthrough*, winter 2012. Retrieved 30 January 2016 from: http://thebreakthrough.org/index.php/journal/past-issues/issue-2/love-your-monsters.

Law, J. (1991). Monsters, machines and sociotechnical relations. In J. Law (ed.), *A Sociology of Monsters: Essays on Power, Technology and Domination* (pp. 1–23). London: Routledge.

Levina, M., and Bui, D.-M. (2013). Toward a comprehensive monster theory in the 21st century. In M. Levina and D.-M. Bui (eds), *Monster Culture in the 21st Century: A Reader* (pp. 1–14). London: Bloomsbury.

Marris, C. (2014). The construction of imaginaries of the public as a threat to synthetic biology. *Science as Culture*, 24, 37–41.

Melucci, A. (1985). The symbolic challenge of contemporary movements. *Social Research*, 52(4), 789–816.

Monsanto (2017). Questions about products of the former Monsanto. *Monsanto*. Retrieved 19 July 2017 from: https://monsanto.com/company/history/articles/former-monsanto-products/.

Nowotny, H. (2014). Engaging with the political imaginaries of science: Near misses and future targets. *Public Understanding of Science*, 23(1), 16–20.

Owen, R., Stilgoe, J., Macnaghten, P., Gorman, M., Fisher, E., and Guston, D. H. (2013). A framework for responsible innovation. In R. Owen, J. Bessant and M. Heintz (eds), *Responsible Innovation: Managing the Responsible Emergence of Science and Innovation in Society* (pp. 27–50). Chichester: John Wiley.

Pratchett, T. (2003). *Monstrous Regiment*. New York: Harper Collins.

Ragonesi, R. (2004). The monster Monsanto. *Times of Malta*, 27 June. Retrieved 29 February 2016 from: www.timesofmalta.com/articles/view/20040627/opinion/the-monster-monsanto.119181.

Reed, M. (2002). Rebels from the crown down: The organic movement's revolt against agricultural biotechnology. *Science as Culture*, 11(4), 481–504.

Riley-Smith, B. (2014). Owen Paterson says 'wicked' green blob protests against GM research kill thousands every day. *Telegraph*, 16 October. Retrieved 2 March 2016 from: www.telegraph.co.uk/news/earth/environment/11166232/Wicked-green-blob-kills-thousands-a-day-through-exploitative-protests-Owen-Paterson-says.html.

Smits, M. (2006). Taming monsters: The cultural domestication of new technology. *Technology in Society*, 28(4), 489–504.

Stavrianakis, A. (2012). Flourishing and Discordance: On Two Modes of Human Science Engagement with Synthetic Biology. PhD thesis, Department of Anthropology, University of California, Berkeley.

Strom, S. (2014). Companies quietly apply biofuel tools to household products. *New York Times*, 31 May. Retrieved 2 March 2016 from: www.nytimes.com/2014/05/31/.business/biofuel-tools-applied-to-household-soaps.html.

Tait, J. (2009). Upstream engagement and the governance of science: The shadow of the genetically modified crops experience in Europe. *EMBO Reports*, 10(1S), 18–22.

Thomas, J. (2014). What-syn-a-name? *Guardian*, 8 July. Retrieved 2 March 2016 from: www.theguardian.com/science/political-science/2014/jul/08/what-syn-a-name.

Thorpe, C., and Gregory, J. (2010). Producing the post-Fordist public: The political economy of public engagement with science. *Science and Culture*, 19(3), 273–301.

Tranter, P. J., and Sharpe, S. (2008). Escaping Monstropolis: Child-friendly cities, peak oil and *Monsters, Inc. Children's Geographies*, 6(3), 295–308.

Turney, J. (2000). *Frankenstein's Footsteps: Science, Genetics and Popular Culture*. New Haven, CT: Yale University Press.

von Schomberg, R. (2013). A vision of responsible research and innovation. In R. Owen, J. Bessant and M. Heintz (eds), *Responsible Innovation: Managing the Responsible Emergence of Science and Innovation in Society* (pp. 51–74). Chichester: John Wiley.

von Schomberg, R. (2014). The quest for the 'right' impacts of science and technology: A framework for responsible research and innovation. In J. van den Hoven, N. Doorn, T. Swierstra, B.-J. Koops and H. Romijn (eds),

Responsible Innovation 1: Innovative Solutions to Global Issues (pp. 33–50). Dordrecht: Springer.

Welsh, I., and Wynne, B. (2013). Science, scientism and imaginaries of publics in the UK: Passive objects, incipient threats. *Science as Culture*, 22(4), 540–566.

Winerip, M. (2013). You call that a tomato? *New York Times*, 24 June. Retrieved 8 July 2016 from: www.nytimes.com/2013/06/24/booming/you-call-that-a-tomato.html.

Wood, D. (1999). Luddites must not block progress in genetics. *Nature*, 397(6716), 201.

Part III

Expertise

9

Expertise

Mark B. Brown

The complex relations among publicity, legitimacy and expertise have long been central to modern science. From the 1660s onward, Robert Boyle and the natural philosophers at the Royal Society legitimated their work in part by portraying it as a distinctly public form of knowledge production. Employing a rhetoric of transparency, they wrote meticulous lab reports in a modest style and performed their experiments in public. They produced expert knowledge both in public and through the public. But their public was largely restricted to elite gentlemen, and they deemed ordinary citizens unqualified and untrustworthy (Golinksi, 2005; Shapin, 1994). Moreover, the founding of the Royal Society coincided not with a democratic revolution, but with the 1660 Restoration of King Charles II.

Today, modern science and democracy are often portrayed as twins, but public acceptance of the former actually preceded the latter by at least a century (depending on how one defines each). Until the nineteenth century, most commentators saw 'democracy' as an unstable and unwise form of government, preferably restricted to the popular branch of systems of 'mixed government' that combined democratic, aristocratic and monarchical elements (Wood, 1992). Liberal elites worried about the corrupting influence of the unwashed masses and the 'tyranny of the majority'. As John Stuart Mill noted in 1859, 'The "people" who exercise the power are not always the same people with those over whom it is exercised' (Mill, 1978 [1859]: 4). The recent increase of xenophobic right-wing populism in both Europe and the United States lends new urgency to such concerns, including questions regarding the publicity and legitimacy of expert knowledge. What aspects of science should be made public and to whom? Will public scrutiny of expertise increase or decrease its legitimacy?

The chapters in this part explore such questions in the context of specific areas of scientific and political controversy: risk governance, biosecurity, climate change, animal research and fisheries management. Democratic citizens have often seen scientific experts as uniquely qualified to speak for the public and the public interest. Many have assumed that science is intrinsically supportive of a wide range of public goods: economic growth, technological progress, national security, cultural enlightenment, informed decision making and, more generally, the public interest itself. But the notion that science has intrinsic value no longer seems widely accepted, and so governments have sought to increase the perceived public value of science by directly involving members of the public in science policy. But public engagement can be messy and unpredictable, and so public officials and other elites typically try to constrain it in various ways.

The first two chapters in this part examine government attempts to engage publics in risk governance. As the authors point out, public officials have usually seen *risk assessment* as a value-free 'scientific' stage of knowledge production, and hence as an unlikely venue for public involvement. In contrast, they have understood *risk management* as a value-laden policy stage, and hence as appropriate for participation by non-experts. In the cases examined here, however, government initiatives to engage the public had the effect of expanding the range of issues and types of participants in only some respects, while restricting them in others.

Chapter 10, by Sarah Hartley and Adam Kokotovich, challenges the standard view of risk governance by disaggregating the risk-assessment stage into different components. The authors argue that public involvement is necessary whenever regulators make choices about values, and they identify a need for value-laden choices in three components of risk assessment: risk-assessment guidelines, the process of conducting risk assessments and the use of scientific studies within risk assessment. This framework allows them to reveal significant implications of the much-discussed 1983 and 1996 reports on risk by the US National Research Council. The first report, the famous 'Red Book', distinguished sharply between risk assessment and risk management, and it did not advocate public involvement in risk assessment. The 1996 report, in contrast, called for involving stakeholders in an 'analytic-deliberative model'. From a public engagement perspective, the second report has widely been seen as an improvement. But Hartley and Kokotovich argue that the 1983 report actually more fully departs from the traditional image of value-free science by highlighting the role of policy 'judgments' in risk assessment. The 1996 report, in contrast, asserts a distinction between value-free expert analysis and value-laden public deliberation. Although the 1996 report expanded opportunities for public engagement, it returned to an outdated view of science as value-free. It thus created new opportunities for public deliberation, while simultaneously constraining its allowable topics. Hartley and Kokotovich then turn to the Codex, an organisation that sets international food-safety standards. They show how regulators failed to address the implicit values in risk assessment, leading to a highly constrained role for public involvement. Overall they show that, as regulatory science has become increasingly open to public involvement, risk-governance procedures have reduced their acknowledgement of values in risk assessment. They conclude by making a nuanced case for public involvement in value-based decisions with regard to specific aspects of risk assessment.

In chapter 11, Judith Tsouvalis examines the UK's emergency response to a deadly fungal disease affecting ash trees, known as Chalara or ash dieback, which gradually spread across Europe during the 1990s and early 2000s. When it reached the UK in 2012, scientists knew little about the disease, and some of what they knew turned out to be wrong. Its arrival thus created an urgent need for government action in the face of scientific uncertainty, leading to an opening of scientific and public discussion through open-source platforms for sharing data and various kinds of citizen science. Tsouvalis shows that these efforts, despite using the language of public engagement, were framed technocratically as narrow matters of 'biosecurity'. A sense of national urgency was fuelled by apocalyptic media images of natural and economic catastrophe. The Government convened an expert task force to provide recommendations for managing the disease, but it only involved community stakeholders once the work was largely complete. The result was a series of narrowly tailored recommendations for risk management and border control. There was no genuine opening of public discussion to address larger issues such as global trade, which, as the task force members themselves recognised, was a key cause of the Chalara outbreak, as well as other plant disease epidemics.

The other two chapters in this part show how controversies over expertise may generate new publics and new views of the public interest. These chapters echo recent work in democratic theory on political representation, which examines how representative claims partly constitute the same publics they represent (Disch, 2015; Saward, 2010). This work does not suggest that representatives create their constituencies out of thin air. When someone makes a claim to represent a constituency, the pre-existing features of the audience – well-established opinions and interests, as well as social and economic conditions – constrain the range of plausible representations and the audience's ability to respond. But what counts as 'representation' in any particular case and the boundaries and identity of the represented are partly determined by the process of representation itself. In a similar vein, these chapters show that those claiming to draw on expertise to represent a particular public or view of the public interest can easily get more than they bargained for, evoking oppositional constituencies beyond their intended audience.

In chapter 13, Warren Pearce and Brigitte Nerlich examine the famous 2006 climate change documentary, *An Inconvenient Truth*, by former US Vice President Al Gore. Climate change is a notoriously difficult issue around which to mobilise publics, because it is a long-term problem that seems much less pressing than most people's everyday concerns. And as a global phenomenon, climate change is invisible to the human senses. Climate change has various local impacts, but the extent to which any particular impact is related to climate change can only be discerned with climate science. Gore's film addressed these challenges by combining personal stories from his life with scenes of him presenting a slide show and talking to audiences about climate science. In some respects, Pearce and Nerlich argue, *An Inconvenient Truth* goes beyond a traditional 'deficit model' of science communication, which sees the key barrier to political action in the public's deficit of scientific knowledge. By making climate change into a personal matter for both himself and the audience, Gore recognised that knowledge alone is not enough. In this respect, they note, the film echoes John Dewey's argument that science education requires aesthetic modes of communication, because 'ideas are effective not as bare ideas but as they have imaginative content and emotional appeal' (Dewey, 1988:169). By presenting himself performing his slideshow in front of various audiences, Gore sought to 'create a public that in turn would continue the performance'. In this respect, Gore's film was successful in 'making the impersonal personal and the invisible visible' (p. 223). At the same time, however, Pearce and Nerlich show that the film embraces a linear model of science education, which assumes that scientific knowledge precedes and compels political action. In this respect, Gore's film promoted a 'hegemonic' representation of climate change, closely linked to his own status as a prominent politician of the US Democratic Party. The film thus became a focal point for an 'inconvenient public' of critics who developed a 'polemic' counter-representation of climate change as either a complete hoax or, at best, an entirely natural phenomenon. Paradoxically, by seeking to make climate change meaningful, the film also politicised the issue in a partisan sense, linking climate change to 'a narrow range of policy options that were anathema to US conservatives' (p. 224). In the future, Pearce and Nerlich argue, climate policy advocates should be more open to engaging with 'uninvited' and 'inconvenient' publics, as well as a wider range of policy options.

Finally, in chapter 13, Sujatha Raman, Pru Hobson-West, Mimi Lam and Kate Millar show how the public interest can be constructed through science controversies. The 'Science Matters' speech by UK Prime Minister Tony Blair in 2002 suggested that scientists and ordinary citizens could work together to pursue the public interest. But Blair also suggested that science inherently promotes a substantive public interest in technological progress, thereby reducing public engagement to a means for promoting this pre-given interest. In contrast, the authors argue that the public interest cannot be known 'in advance of concrete efforts' to engage actual citizens (p. 233). They also note that research on science controversies typically focuses on procedural questions of inclusion and exclusion, which can result in the analysis unwittingly reproducing a 'fragmented, individualised version of the public' (p. 240), as well as a neglect of substantive questions of the public interest. Raman and colleagues go beyond a procedural and individualised approach by exploring struggles between dominant and subordinate advocacy networks over the meaning of the public interest, focusing on cases involving animal research in the UK and fisheries in Canada. In the case of animal research, public officials portrayed those opposed to such research as opponents of the public interest. They presented opinion polls supportive of animal research as a more authentic representation of the public. In most respects, proponents of animal research have so far maintained the upper hand. In the fisheries case, in contrast, a Canadian First Nation community won a victory against the Government. The Haida Nation opposed the Government's fisheries policy because the policy threatened not only its culture and traditional values, but also the local ecology and the larger public interest. In a court case the judge concluded that the Haida Nation, rather than the Government, had the stronger claim to the public interest. With these cases the authors show that public engagement often does more than just 'open up' science to public scrutiny. It may also lead to 'renegotiating the substantive question of what is in the public interest' (p. 246).

As these four chapters indicate, the relation of scientific expertise, the public interest and democratic legitimacy is today highly contested. In some respects, this should not be surprising, because democratic legitimacy involves an irresolvable tension between majority rule and public justification. It requires that the general public accept the *majority* of the people as standing for the *entire* people (Rosanvallon, 2011). Popular elections once offered a relatively effective means of securing such acceptance, but democracies have not been able to rely solely on elections as a source of legitimacy since the late nineteenth century. With the adoption of universal voting rights, citizens increasingly questioned the notion that majority rule alone could guarantee governmental virtue and competence. A potential remedy appeared in the rise of bureaucratic expertise and the administrative state, which promised a substantive form of legitimacy to complement procedural legitimacy through popular elections. But since at least the 1980s, both elections and expertise have faced widespread public scepticism. Many people do not trust experts to remain objective, and 'the people' are fragmented into a plethora of ever-shifting interest and identity groups. In the United States, for example, millions of citizens questioned the basic legitimacy of elected presidents George W. Bush and Barack Obama, driven by concerns about substantive competence and group identity, respectively. This widespread epistemic scepticism and social fragmentation has led to a 'radical pluralisation of the forms of legitimacy' (Rosanvallon, 2011: 8). Government commissions, expert committees and lay deliberative bodies all seek to shore up the faltering legitimacy of democratic governments. The emergence of multiple forms of legitimacy means that no single expert recommendation or political institution, and especially no populist demagogue, should be taken as the authentic voice of the people. Just as technocracy reduces democracy to the truth claims of science, populism reduces it to a monolithic representation of the will of the people.

Avoiding technocracy without fostering populism is a key challenge of our time. To what extent should lay citizens become involved in shaping expert recommendations? What are the most promising means of communicating expert knowledge to diverse constituencies? How can we best involve experts in political contests over the meaning and content of the public interest? The chapters in this part do not seek definitive answers to these questions, but they offer valuable resources for thinking about them.

References

Dewey, J. (1988). Freedom and culture. In *The Later Works*, vol. 13, *1938–1939*, ed. by J. A. Boydston (pp. 63–188). Carbondale and Edwardsville, IL: Southern Illinois University Press.

Disch, L. (2015). The 'constructivist turn' in democratic representation: A normative dead end? *Constellations*, 22(4), 487–499.

Golinski, J. (2005). *Making Natural Knowledge: Constructivism and the History of Science*. Chicago: University of Chicago Press.

Mill, J. S. (1978 [1859]). *On Liberty*, ed. by E. Rapaport (originally published 1859). Indianapolis, IN: Hackett.

Rosanvallon, P. (2011). *Democratic Legitimacy: Impartiality, Reflexivity, Proximity*, trans. by A. Goldhammer. Princeton, NJ: Princeton University Press.

Saward, M. (2010). *The Representative Claim*. Oxford: Oxford University Press.

Shapin, S. (1994). *A Social History of Truth: Civility and Science in Seventeenth Century England*. Chicago: University of Chicago Press.

Wood, G. S. (1992). Democracy and the American Revolution. In J. Dunn (ed.), *Democracy: The Unfinished Journey, 508 BC to AD 1993* (pp. 95–105). Oxford: Oxford University Press.

10

Disentangling risk assessment: new roles for experts and publics

Sarah Hartley, Adam Kokotovich

Risk assessment is an important stage of risk governance, alongside risk characterisation, risk evaluation and risk management. A burgeoning literature on public involvement in risk governance and science-based policymaking more broadly has developed in response to tensions in governing environmental risk, particularly the environmental risks posed by emerging technologies (Irwin, 2014; Levidow, 2007; Renn and Schweizer, 2009; Rothstein, 2013; Wynne, 2006). However, there is relatively little investigation of public involvement in the specific stage of risk assessment, despite increased demands for such involvement (Borrás et al., 2007; Hartley, 2016; Millstone, 2009; Shepherd, 2008). European and North American regulatory agencies have a statutory obligation to involve the public in risk governance, and in recent years many have opened up the traditionally scientific domain of risk assessment to public input through online consultations. In addition, international bodies have created opportunities to engage a broader range of experts and stakeholders. However, there is evidence that regulatory agencies and international organisations are not meeting their statutory obligations, falling short of their own guidelines in practice (Dreyer and Renn, 2014; Hartley, 2016; Herwig, 2014).

We argue that public involvement in risk assessment is not reaching its full potential owing to a considerable lack of clarity in the literature and in practice about which publics should be involved in risk assessment and at what point they should be involved. Much of the risk-governance literature examining public involvement fails to disentangle adequately the process of risk assessment when examining questions about who to involve, when to involve them, and why. Risk assessment is not a single stage of risk governance that can simply be made participatory; rather, it is a process with different components that need to be considered individually when determining how and why to open risk assessment to publics. Furthermore, for the potential of public involvement to be fully realised, a particular understanding of risk assessment is necessary – one that is detailed and that recognises inherent value judgements. Conflating the different aspects of risk assessment and the different types of participation makes opening risk assessment to publics seem unreasonable and risks the legitimacy of regulatory agencies.

We draw on the theoretical, prescriptive and empirical literature to disentangle risk assessment for governing human health and environmental risks of emerging technologies. This disentanglement begins with an examination of values in risk assessment and restates the case for public involvement when value choices are to be made. First, we argue that effective and legitimate public involvement is dependent upon the degree to which value judgements are acknowledged in the different components of risk assessment. Second, we explore variations in the prescription literatures of the National Research Council (NRC) in the USA, and the international organisation the Codex Alimentarius Commission (CAC, or 'Codex'). Third, we examine the way in which risk assessment is disentangled in practice through the case study of the European Food Safety Authority (EFSA). Finally, we draw on these findings to reassemble public involvement in risk assessment, making clear who should be involved, where and, importantly, why.

Disentangling values and risk assessment: the need for public involvement

Risk governance involves a number of stages, and a plethora of different models exist. The delineation of the various stages depends upon the degree to which the 'scientific' stage can be separated from the 'policy' stage. In general terms, these models do separate the scientific stage (risk assessment) from the policy stage (risk management). However, there is considerable evidence to suggest these stages are not separated in practice (Millstone, 2009). Much of the risk-governance literature that addresses public involvement fails to disentangle risk assessment from risk management adequately; for example, Renn and Schweizer (2009) suggest there is a default assumption that public involvement should occur in risk management. Kaliarnta et al. (2014) examine stakeholder involvement in risk governance but focus on questions concerning who should participate and why, how much they should participate, and what the participation should address, but they do not tease apart the stages of assessment and management to ask where public involvement should take place.

Where the literature has distinguished the stages of risk governance to examine where public involvement might be best utilised, risk assessment remains an epistemic stage that is seen as insulated from values – with consequences for how public involvement is envisioned in risk assessment. For example, Dreyer and Renn (2014) lay out four stages in risk governance –framing (design discourse), appraisal (epistemic discourse), evaluation (reflective discourse) and management (practical discourse). While they argue that publics can make a contribution to the epistemic discourse, this discourse does not involve the discussion of value choices, which are dealt with in the evaluation and management stages. Therefore, publics, who are often asked to comment on published risk-assessment documents in online consultations, are restricted in terms of the types of input they are able to provide during consultation. Consequently, the value choices inherent in risk assessment are not open to public scrutiny and publics are able to comment only on the scientific aspects of risk assessment, and only on scientific terms.

A key aspect of whether and how to involve publics in risk assessment is based on how we understand the role of values within risk assessment. First, and at the broadest level, choosing to use risk assessment to inform decision making is itself a value-based decision. To frame an issue in terms of risk and risk assessment will privilege certain actors and marginalise other possible ways of understanding that issue (Jasanoff, 1999). Second, if risk assessment itself is understood as an objective scientific process external to value judgements, there is little role for public involvement other than, perhaps, for expert stakeholders to ensure the science is completed correctly (Jasanoff, 1987). Once the role of values in risk assessment is acknowledged and reflected upon, the need for public involvement is strengthened.

There is an extensive body of literature that demonstrates the relevance of values throughout the risk-assessment process (Kokotovich,

2014; Meghani, 2009). Challenging the notion that values can be confined to risk management, this scholarship explores how normative values influence all aspects of risk assessment. This work shows that such judgements have consequences and thus need to be taken seriously, including by opening them to reflection and public involvement. Here we identify and review the value-based nature of three key components of risk assessment: the guidelines that shape risk assessment, the conduct of risk assessment and the science used in risk assessment.

Risk-assessment guidelines

Guidelines establish the steps to follow when conducting a risk assessment and provide assistance both to applicants preparing risk assessments and to risk assessors conducting them. Thus, they incorporate value judgements about the scope of future risk assessments, including what falls inside the scope of risk assessment; what counts as evidence, how much evidence is needed and how it should be interpreted; and how uncertainty should be addressed. There is a growing realisation of the importance of guidelines. Kokotovich (2014), for example, studied two competing sets of guidelines for assessing the risks to non-target organisms from insect-resistant genetically modified plants, and found their divergent foundational value judgements resulted in recommending different processes for risk assessment, and different potential outcomes. These judgements involved the adequacy of substantial equivalence testing and what species needed to be tested, together with the (un)importance of assessing indirect effects, and they resulted in the guidelines calling for different kinds of scientific studies to be completed to inform the risk assessment. Millstone et al. (2008) show that differences in guidelines account for transatlantic trade conflicts such as those that arose over beef hormones, recombinant bovine somatotrophin and genetically modified maize.

Conducting risk assessment: problem formulation, analysis and risk characterisation

Conducting risk assessment is a process that includes problem formulation, exposure and effects analysis, and the characterisation of risk. Many of the decisions in risk assessment that are acknowledged as value based and that have been opened to public involvement occur in the formulation of the problem (Environmental Protection Agency, 1998; Nelson et al., 2007). Problem formulation is the initial step of the risk assessment that determines the assessment endpoints, the conceptual model linking the stressors to the assessment endpoints, and an analysis plan. This step is widely seen as the place where values most explicitly enter the risk-assessment process. Authors such as Thompson (2003) and Jensen et al. (2003) have revealed how decisions taken in problem formulation, such as identifying the specific hazard to be assessed and determining the time and spatial scale, are value based. These decisions alter the scope of the risk assessment in ways that can influence the ultimate characterisation of the risk. Problem formulation is where values and public involvement are often acknowledged and allowed, and this is also where they are classically confined. Similar to the distinction between risk assessment and risk management within the classical notion of risk governance, problem formulation sets up a dichotomy between science and values. In this understanding, values exist in the formulation of the problem, while the scientific analysis phase remains free from values. Problem formulation does, however, stand apart from the rest of risk assessment owing to the type of value judgements that need to be made. Many of these judgements involve explicit value-based, non-technical judgements that do not require technical expertise from contributors.

While the discussion of values in conducting risk assessment normally begins and ends with problem formulation, the analysis and risk-characterisation steps also contain value judgements. For example, identifying and synthesising relevant scientific studies and addressing uncertainty all involve value judgements that can influence the overall assessment of risk (Meyer, 2011; Winickoff et al., 2005). The differences in how these value judgements are addressed contribute to the reason why different regulatory bodies can arrive at differing assessments of risk (Wickson and Wynne, 2012). The value judgements in these steps require a greater degree of technical expertise than those at the stage of problem formulation.

Scientific studies used in risk assessments

This component of risk assessment is rarely considered distinctly. However, the scientific studies used in a risk assessment are also influenced by value judgements and therefore should not escape scrutiny (Elliott, 2012; Holifield, 2009). Scientific studies influence the ultimate characterisation of risk, yet they themselves can be influenced by the different parts of a risk assessment. Both the development of risk-assessment guidelines and the conducting of a risk assessment involve value choices over what scientific studies are relevant (Kokotovich, 2014). Risk-assessment guidelines can influence how scientific studies are completed by, for example, calling for the use of surrogate species or local species in laboratory testing (Hilbeck et al., 2011). There can also be value judgements in the design and conduct of scientific studies that go beyond those stipulated in risk-assessment guidelines and the conduct of a risk assessment. Scientific studies used in a risk assessment depend on the often subtle value judgements that inform them (Elliott, 2012; Holifield, 2009). Elliot calls attention to the notion of 'selective ignorance', or the 'wide range of often subtle research choices or "value judgements" that lead to the collection of some forms of knowledge rather than others' (2012: 331), claiming these judgements will influence what knowledge is available to inform decision making or, in our case, risk assessment.

The existence of value judgements in these three components of risk assessment draws attention to the actors making those value judgements. Who is making them and who should do so? The recognition of these value choices has fuelled the call for democratic accountability and public involvement in risk assessment, which has traditionally been seen as an expert domain (Hartley, 2016).

Prescribing the treatment of values and publics in risk assessment

In reviewing key examples of the existing prescriptive risk-assessment literature, specifically documents from the NRC and Codex, we show how values are acknowledged and public involvement is proposed by the organisations that prescribe risk assessment (NRC, 1983; Stern and Fineberg, 1996). In comparing the 1983 and 1996 NRC reports, we argue that it is the 1983 report that acknowledges the role of values in risk assessment in a more detailed, nuanced and potentially productive way. This is true even though it calls for a separation of risk assessment and risk management, and the 1996 report calls for broader public involvement in risk governance and for an integration of assessment and management.

The NRC is part of the National Academies of Science, a private, non-profit, self-perpetuating society of distinguished scholars engaged in scientific and engineering research. It advises the US Federal Government on scientific and technical matters and has published several prominent reports on risk assessment (NRC, 1983, 2009; Stern and Fineberg, 1996). These NRC reports have influenced regulatory risk assessment in the USA and internationally (Suter, 2008), and they show how values and the role of public involvement are acknowledged in risk assessment.

Risk Assessment in the Federal Government: Managing the Process was one of the first major reports on risk assessment. This report has become known as the 'Red Book' because of its red cover. It supported the clear separation of risk assessment from risk management to help establish the credibility of risk assessment (NRC, 1983), but, at the same time, recognised the value judgements that are entangled in risk assessment. For example, the NRC states:

> If risk assessment as practiced by the regulatory agencies were pure science, perhaps an organizational separation [between risk assessment and risk management] could effectively sharpen the distinction between science and policy in risk assessment and regulatory decision-making. However, many of the analytic choices made throughout the risk assessment process require individual judgments that are based on both scientific and policy considerations. (NRC, 1983: 143)

The NRC refers to value judgements as policy judgements, and introduces the concept of risk-assessment policy to refer to them in conducting risk assessments. The value judgements inherent in risk assessment are seen as being different in character from the value judgements that exist in risk-management decisions. Making a clear distinction between the types of value choices present in risk assessment and risk management, the NRC describes how to distinguish between scientific and value judgements in risk assessment, which it notes is a difficult task. It recommends the development of guidelines, which it defines as 'the principles followed by risk assessors in interpreting and reaching judgments based on scientific data' (NRC, 1983: 51). These guidelines help the risk assessor in conducting future risk assessments and are similar to the risk-assessment guidelines that we describe above. However, there is no mention of the scientific studies used in a risk assessment or the values-based nature of the research.

In this 1983 report the NRC argues that value judgements in risk assessment are best made by risk assessors and there is no suggested role for public involvement, even though the report recognises the implications of such judgements. However, the NRC recommends the involvement of experts from a wide range of scientific disciplines in the development of guidelines. Overall, then, this report provides an understanding of risk assessment in which the role of values in risk assessment is acknowledged. These judgements are to be addressed by expert risk assessors who will follow risk-assessment guidelines that have been developed by a broad range of experts in advance of an individual risk assessment.

In 1996 the NRC published the report *Understanding Risk: Informing Decisions in a Democratic Society*, which proposes an analytic–deliberative approach to risk governance, one where 'deliberation frames analysis [and] analysis informs deliberation' throughout the entire risk-governance process (Stern and Fineberg, 1996: 6). In contrast to its 1983 report, the NRC extends its thinking about the role of values in risk governance and the way in which they should be dealt with. It proposes broad involvement in risk governance by experts, decision makers, and interested and affected parties. Yet the distinction between analysis and deliberation makes clear the separation between analysis, an epistemic stage that is the domain of experts, and deliberation, which can be opened up to non-experts. Analysis uses rigorous, replicable methods developed by experts to arrive at answers to factual questions. Deliberation includes value-based decisions with a focus on how issues are framed and what questions need to be answered. It uses processes such as discussion, reflection and persuasion to communicate, raise and collectively consider issues, increase understanding and arrive at substantive decisions (Stern and Fineberg, 1996: 20).

Interested and affected parties can influence what analysis is called for, but they play less of a role in actually influencing that science and the risk-assessment process. Their direct involvement may be possible when they have specialised or local knowledge that can help inform the analysis. While they are not brought explicitly into the analysis, they are seen as having a role in at least checking it:

> Participation is important to help ask the right questions of the science, check the plausibility of assumptions, and ensure that any synthesis is both balanced and informative. The more likely it is that the science will be criticized on the basis of its underlying assumptions or alleged omissions, the more important participation is likely to be in a risk decision process. (Stern and Fineberg, 1996: 132)

This approach points to the values-based nature of such judgements, while not explicitly calling for participation in the development of risk-assessment guidelines or in the conduct of a risk assessment where the assumptions and synthesis are determined. Overall, this report takes a bird's-eye view of risk governance without the nuanced attempt to disentangle the process of risk assessment that the 1983 report contained. Rather, it proposes the conceptual separation of the analytical stage of governance from its deliberative stage. Therefore, risk assessment has become a single stage in an approach to risk governance which is free of values and the domain of expert risk assessors. Public involvement is confined to risk management.

Codex is an international organisation established in 1963 by the Food and Agriculture Organization and the World Health Organization (WHO) to develop harmonised science-based international food standards and risk-assessment procedures in order to protect consumer health and ensure fair international trade (Büthe and Harris, 2011). It is a heavily expert-led organisation and exclusively science based, although this reliance on science and scientific experts to the exclusion of non-scientific factors, alternative experts and mechanisms for public involvement has been strongly criticised (Foster, 2008; Herwig, 2014; Peel, 2010).

Codex constitutes the next phase of thinking about risk assessment, developing the concept of 'risk-assessment policy' which is similar to the risk-assessment guidelines that we described above. According to Codex, risk-assessment policy establishes the risk-assessment framework, and is defined as 'documented guidelines on the choice of options and associated judgments for their application at appropriate decision points in the risk assessment such that the scientific integrity of the process in maintained' (CAC, 2013: 114). These judgements include decisions about the scope of future risk assessments, the type and amount of evidence needed, the interpretation of the evidence, and the treatment of uncertainty. The concept departs from the NRC's reports by stipulating that it is the responsibility of risk managers, not risk assessors, to develop a risk-assessment policy in consultation with all interested parties (CAC, 2013). In 2007, Codex committed its 186 members (including the EU and the USA) to develop explicit risk-assessment policies through the formal adoption of the *Working Principles for Risk Analysis for Food Safety for Application by Governments* (CAC, 2007).

Similar to the NRC in its 1996 report, Codex acknowledges the value judgements in risk-assessment guidelines, yet it maintains the clear distinction between risk assessment and risk management. Risk-assessment guidelines have been carved off from risk assessment and placed under the risk-management phase of risk governance. However, the more nuanced discussion about value judgements in risk assessment present in the NRC's 1983 report has been pushed aside in favour of a clear and convenient separation between facts and values. Risk assessment is now seen to be an exclusively science-based and objective exercise to be conducted by risk assessors (Herwig, 2014). Risk management is the stage of risk governance where values are acknowledged and where the public should be involved. Next, we explore these prescriptions and the tensions that arise in the practical application of guidelines through a case study.

The treatment of values and publics in risk assessment in practice

The EFSA provides a useful case study to examine the way in which risk assessors disentangle risk assessment and involve the public in practice. The EFSA gives independent scientific advice to the European Commission (EC) on matters related to food safety, and has responsibility for risk assessment. Risk assessment is defined in the EFSA's founding regulation as 'a scientifically based process consisting of four steps: hazard identification, hazard characterization, exposure assessment and risk characterization' (EC, 2002: 11). The EFSA develops guidance documents, which are risk-assessment guidelines establishing the principles, procedures and approaches in risk assessment as well as specifying data requirements and the handling of uncertainty (Hartley, 2016; Vos and Wendler, 2006). Applicants conduct risk assessment in line with the EFSA's guidance documents and then the EFSA reviews the applications and publishes a scientific opinion (the output of an individual risk assessment). It is then the task of the EC and member states (risk managers) to make the decision on whether to approve the product or process under scrutiny.

The EFSA relies heavily on independent external scientific experts in the development of its scientific outputs. These experts sit on standing panels and are called upon to sit on ad hoc working groups. In addition, the EFSA has a statutory obligation to engage with publics (EC, 2002: Article 42). To meet this obligation, the EFSA holds public consultations on its scientific outputs, particularly its guidance documents and scientific opinions. Public involvement is guided by an internal policy. The EFSA's approach to public consultations on scientific outputs defines publics as 'the non-institutional stakeholders, which include academics, NGOs, industry and all other potentially interested and affected parties' (EFSA, n.d.: 3). The EFSA's motivation for public consultation in risk assessment is driven by the goals of both transparency and scientific excellence. Public consultations open up the EFSA's processes and decisions to public scrutiny and they also allow external input from publics to enhance the scientific quality of the risk assessment by ensuring clarity and completeness (EFSA, n.d.: 3). The EFSA's policy on consultations allows it to launch a public consultation at three stages: (1) at the start, to define the scope and major principles; (2) at a preliminary stage, to seek information, data, views and sources available on a specific topic; and (3) at the end, to ensure the clarity, completeness and soundness of the draft scientific output (EFSA, n.d.). However, the EFSA has yet to hold a consultation at the first stage to define the scope of a risk assessment. In practice, publics are typically given two months to comment on a draft scientific output (developed by experts) through the EFSA's website (Hartley, 2016).

The EFSA does not acknowledge that value judgements are made in the development of its guidance documents or scientific opinions (Klintman and Kronsell, 2010). Independent experts on the EFSA's Genetically Modified Organisms (GMO) panel have made it clear that they do not acknowledge or engage in the matter of implicit values, instead insisting that the EFSA's risk assessment is a scientific process and value judgements occur at the risk-management stage of risk governance and are the responsibility of the EC and member states (Perry et al., 2012; Wickson and Wynne, 2012). Further, the

EFSA officials and scientific panels do not recognise that guidance documents are risk-assessment policies, as defined by Codex, or that it is the EC's responsibility to develop them (Hartley, 2016). Guidance documents are treated as scientific outputs free of value judgements. However, despite the legal distinction between risk assessment and risk management and the EFSA's insistence that risk assessment is value free, in practice the distinction is blurred (Tai, 2010).

The institutional denial of value judgements in risk assessment has significant implications for the EFSA's public consultations. First, it means that the EFSA's public consultations are 'science based' and publics are allowed only to provide comments related to the science of risk assessment. For example, when the EFSA consulted the public in the development of its guidance documents on the environmental risk assessment of genetically modified animals in 2013, it informed potential participants: 'The EFSA GMO Panel considered all scientifically relevant comments from the public when finalising the present document. [It] did not consider issues related to risk management (e.g. traceability, labelling, coexistence). Ethical and socio-economic issues are also outside the remit of the EFSA GMO Panel' (EFSA, 2013: 6). However, guidance documents are risk-assessment policies and the EU's commitment to Codex rules requires the EC to develop them. Hartley (2016) has described these guidance documents as policies masquerading as science.

The second implication of the institutional denial of values in risk assessment is that it reinforces the authority of experts, and publics have minimal opportunity for influence through the consultation. Hartley (2016) argues that the public consultations have a minimal impact on the EFSA's scientific outputs owing to the expert-led nature of the process and the unjustified restrictions placed on public involvement. Gaskell et al. (2007) characterise the EFSA's public consultation approach as a 'sound science' type of public dialogue, where the EFSA listens to the public only in terms of its own expert definition of the problem and the possible solutions. Although the EFSA makes public the results of the consultation exercises and its response, which shows how the results of the consultation exercise are used, publics' views are heard only in so far as publics talk in terms of the EFSA's scientific remit.

Overall, the EFSA has responsibility for developing risk-assessment guidelines, conducting risk assessment and determining the scientific studies used in a risk assessment in the EU's broader risk-governance framework. Each of these components of risk assessment is seen to be epistemic and is conducted by the EFSA's independent experts. Publics are involved as a means to improve the quality of the science and to make the process of risk assessment transparent. However, in practice, the institutional denial of values in risk assessment means that it is independent experts who determine the values-based decisions, and these experts are not democratically accountable. Public involvement is restricted to matters of science and the value judgements made by experts are hidden from public scrutiny.

New roles for public involvement in risk assessment

The academic literature presents compelling evidence of the existence of values in risk assessment and makes a convincing case that risk assessment has different component parts and should not be considered a homogeneous stage in risk governance. The prescriptive literature of the NRC and Codex demonstrate the difficulty in disentangling risk assessment in practice, showing that since the late 1980s there has been a growing reluctance to take a nuanced approach to addressing values used in risk assessment. Ironically, this closing down of values has been happening at the same time that risk assessment has been opened up to publics. At present, there is no harmonised approach to acknowledging or handling values in risk assessment, or to thinking about how risk assessment should be disentangled. The case of the EFSA reveals that the values in risk assessment are denied in practice and that the different component parts (risk-assessment guidelines, conducting of risk assessment and scientific studies used within a risk assessment) are seen as a single stage of risk governance. This practice of risk assessment has serious implications for public involvement.

The lack of clarity about which publics should be involved in risk assessment and at what point they should be involved means that public engagement in risk assessment is not reaching its full potential. To address this lack of clarity, we have disentangled risk assessment into three components: (1) risk-assessment guidelines, (2) conducting risk assessment and (3) scientific studies used in a risk assessment. Table 10.1 outlines these risk-assessment components. The types of

Disentangling risk assessment

Table 10.1 A framework for public involvement in risk assessment

Risk-assessment components		Task at hand	Type of public to be involved
Risk-assessment guidelines		Establishing the risk-assessment framework	Broad range of alternative experts and publics
Conducting risk assessment	Problem formulation	Defining the scope of and plan for a risk assessment	Broad range of alternative experts and publics
	Analysis and characterisation of risk	Exposure and effects analysis, including selecting and synthesising relevant studies	Alternative experts
Scientific studies for risk assessment		Designing and conducting scientific studies which are drawn upon during risk assessments	Alternative experts

publics to engage with risk assessment will depend upon the component of risk assessment.

We make a practical distinction for the purposes of this argument between alternative experts and general publics, recognising that this distinction may be a false distinction at times. Alternative experts need to be sought out by risk assessors for their expert knowledge, which expands the existing range of expertise. These experts will be able to address the epistemic questions raised in risk assessment and may come from a broader range of academic disciplines, including the natural, engineering and social sciences. Alternative experts may also come from sector-specific policy communities outside the academy such as civil society, policymakers and government risk assessors. Alternative experts may be brought into existing committees, working groups and panels and work alongside risk assessors. On their part, general publics will be self-selected in open and transparent engagement mechanisms in order to allow stakeholder groups and individuals access to information and provide them with the opportunity to contribute to values-based questions. General publics cannot be restricted to answering epistemic questions.

Mirroring developments in the public-engagement literature, there is increasing recognition in the risk-assessment literature of the role of public involvement in contributing substantively to risk assessment and providing transparency (Klintman and Kronsell, 2010). Indeed, the EFSA makes it clear that its public consultations are designed to satisfy both these goals. Therefore, the goal of public involvement in risk assessment is democratic and epistemic legitimacy. However, because the judgements in risk assessment are both science- and values-based in nature, epistemic legitimacy requires democratic legitimacy. There is a need, then, to involve the appropriate publics in the specific component being addressed. Because of the types of value judgements that exist in the development of risk-assessment guidelines and in the problem-formulation stage of risk assessment, including those that do not involve technical expertise, public involvement needs to include both alternative experts and publics more broadly. During the analysis and characterisation of risk and for scientific studies, it is important to open up to alternative experts who hold enough expertise to reflect substantively on the relevant values-based questions.

Conclusion

Peel (2010) suggests one of the crucial issues facing risk assessment and governance is related to the way in which facts and values are addressed: '[It is] not whether science or values should triumph, but rather how scientific and non-scientific inputs might be blended in risk assessment in different settings to ensure a broadly acceptable balance of credibility and legitimacy concerns' (Peel, 2010: 10). We argue that in order to satisfy epistemic and democratic legitimacy, the different features of risk assessment must be disentangled to lay bare the various component parts, and that different publics need to be involved depending on the types of questions asked in each component.

This chapter highlights the tensions between evidence, prescription and practice in risk assessment which complicate efforts to involve publics. However, public involvement in risk assessment presents a significant opportunity to debate the value judgements that exist in the various components of risk assessment. Indeed, it is precisely these implicit value judgements that present the strongest argument for public involvement (Finardi et al., 2012). In contrast, denying that values exist in risk assessment, relying on a narrow range of expertise and limiting public input to epistemic matters imposes a certain set of values made by a narrow range of experts that are insulated from public scrutiny and debate. This institutional denial of the implicit values in risk assessment results in public frustration and lack of trust in regulatory authorities (Hartley, 2016; Wynne, 2006).

References

Borrás, S., Koutalakis, C., and Wendler, F. (2007). European agencies and input legitimacy: EFSA, EMeA and EPO in the post-delegation phase. *Journal of European Integration*, 29(5), 583–600.

Büthe, T., and Harris, N. (2011). Codex Alimentarius Commission. In T. Hale and D. Held (eds), *Handbook of Transnational Governance: Institutions and Innovations* (pp. 219–228). Cambridge, MA: Polity Press.

CAC (2007). *Working Principles for Risk Analysis for Food Safety for Application by Governments*. CAC/GL 62-2007. Rome: Food and Agriculture Organization (FAO).

CAC (2013). *Procedural Manual of the Codex Alimentarius Commission*, 21st edition. Joint FAO/WHO Food Standards Programme. Rome: FAO.

Dreyer, M., and Renn, O. (2014). EFSA stakeholder and public involvement policy and practice: A risk governance perspective. In A. Alemanno and S. Gabbi (eds), *New Directions in EU Food Law and Policy: Ten Years of European Food Safety Authority*. Farnham: Ashgate.

EC (2002). Regulation (EC) No 178/2002 of the European Parliament and of the Council of 28 January 2002 laying down the general principles and requirements of food law, establishing the European Food Safety Authority and laying down procedures in matters of food safety. *Official Journal of the European Communities*, 45(L 31), 1–24.

EFSA (2013). *Guidance on the Environmental Risk Assessment of Genetically Modified Animals*. Scientific Opinion. Parma, Italy: EFSA.

EFSA (n.d.). *EFSA's Approach on Public Consultations on Scientific Outputs*. Parma, Italy: EFSA.

Elliott, K. C. (2012). Selective ignorance and agricultural research. *Science, Technology, and Human Values*, 38(3), 328–350.

Environmental Protection Agency (EPA) (1998). *Guidelines for Ecological Risk Assessment*. Washington, DC: EPA.

Finardi, C., Pellegrini, G., and Rowe, G. (2012). Food safety issues: From enlightened elitism towards deliberative democracy? *Food Policy*, 37, 427–438.

Foster, C. E. (2008). Public opinion and the interpretation of the World Trade Organisation's agreement on sanitary and phytosanitary measures, *Journal of International Economic Law*, 11(2), 427–458.

Gaskell, G., Kronberger, N., Fischler, C., Hampel, J., and Lassen, J. (2007). *Consumer Perceptions of Food Products from Cloned Animals. A Social Scientific Perspective*. Parma, Italy: EFSA.

Hartley, S. (2016). Policy masquerading as science: An examination of non-state actor involvement in risk assessment policy for genetically modified animals in the EU. *Journal of European Public Policy*, 23(2), 276–295.

Herwig, A. (2014). Health risks, experts and decision-making within the SPS Agreement and the Codex Alimentarius. In M. Ambrus, K. Arts, E. Hey and H. Raulus (eds), *The Role of 'Experts' in International and European Decision-making Processes: Advisors, Decision-Makers or Irrelevant Actors* (pp. 194–215). Cambridge: Cambridge University Press.

Hilbeck, A., Meier, M., Römbke, J., Jänsch, S., Teichmann, H., and Tappeser, B. (2011). Environmental risk assessment of genetically modified plants: Concepts and controversies. *Environmental Sciences Europe*, 23(13), 1–12.

Holifield, R. (2009). How to speak for aquifers and people at the same time: Environmental justice and counter-network formation at a hazardous waste site. *Geoforum*, 40(3), 363–372.

Irwin, A. (2014). From deficit to democracy (re-visited). *Public Understanding of Science*, 23(1), 71–76.

Jasanoff, S. (1987). EPA's regulation of daminozide: Unscrambling the messages of risk. *Science, Technology, and Human Values*, 12(3–4), 116–124.

Jasanoff, S. (1999). The songlines of risk. *Environmental Values*, 8, 135–152.

Jensen, K. K., Gamborg, C., Madsen, K. H., Jorgensen, R. B., von Krauss, M. K., Folker, A. P., and Sandoe, P. (2003). Making the EU 'risk window' transparent: The normative foundations of the environmental risk assessment of GMOs. *Environmental Biosafety Research*, 2(3), 161–171.

Kaliarnta, S., Hage, M., and Roeser, S. (2014). Involving stakeholders in risk governance: The importance of expertise, trust and moral emotions. In M. B. A. van Asselt, M. Everson and E. Vos (eds), *Trade, Health and the Environment: The European Union Put to the Test* (pp. 235–253). Abingdon: Routledge.

Klintman, M., and Kronsell, A. (2010). Challenges to legitimacy in food safety governance? The case of the EFSA. *European Integration*, 32(3), 309–327.

Kokotovich, A. (2014). Delimiting the study of risk: Exploring values and judgments in conflicting GMO ecological risk assessment guidelines. In *Contesting Risk: Science, Governance and the Future of Plant Genetic Engineering* (pp. 13–67). PhD thesis, University of Minnesota.

Levidow, L. (2007). European public participation as risk governance: Enhancing democratic accountability for agbiotech policy? *East Asian Science, Technology and Society*, 1(1), 19–51.

Meghani, Z. (2009). The US' Food and Drug Administration, normativity of risk assessment, GMOs, and American democracy. *Journal of Agricultural and Environmental Ethics*, 22(2), 125–139.

Meyer, H. (2011). Systemic risks of genetically modified crops: The need for new approaches to risk assessment. *Environmental Sciences Europe*, 23(7), 1–11.

Millstone, E. (2009). Science, risk and governance: Radical rhetorics and the realities of reform in food safety governance. *Research Policy*, 38, 624–636.

Millstone, E., van Zwanenberg, P., Levidow, L., Spök, A., Hirakawa, H., and Matsuo, M. (2008). *Risk-Assessment Policies: Differences across Jurisdictions*. Brussels: EC Joint Research Centre.

Nelson, K. C., Basiao, Z., Cooper, A. M., Dey, M., Fonticiella, D., Hernandez, M. L., Kunawasen, S., et al. (2007). Problem formulation and options assessment: Science-guided deliberation in environmental risk assessment of transgenic fish. In A. Kapuscinski, K. Hayes, S. Li and G. Dana (eds), *Environmental Risk Assessment of Genetically Modified Organisms*, vol. 3, *Methodologies for Transgenic Fish* (pp. 29–60). Cambridge, MA: Centre for Agriculture and Biosciences International.

NRC (1983). *Risk Assessment in the Federal Government: Managing the Process*. Washington, DC: National Academies Press.

NRC (2009). *Science and Decisions: Advancing Risk Assessment*. Washington, DC: National Academies Press.

Peel, J. (2010). *Science and Risk Regulation in International Law*. Cambridge: Cambridge University Press.

Perry, J. N., Arpaia, S., Bartsch, D., Kiss, J., Messéan, A., Nuti, M., Sweet, J. B., et al. (2012). Response to 'the anglerfish deception'. *EMBO Reports*, 13, 481–482.

Renn, O., and Schweizer, P.-J. (2009). Inclusive risk governance: Concepts and application to environmental policy making. *Environmental Policy and Governance*, 19(3), 174–185.

Rothstein, H. (2013). Domesticating participation: Participation and the institutional rationalities of science-based policy-making in the UK Food Standards Agency. *Journal of Risk Research*, 16(6), 771–790.

Shepherd, R. (2008). Involving the public and stakeholders in the evaluation of food risks. *Trends in Food Science and Technology*, 19(5), 234–239.

Stern, P. C., and Fineberg, H. V. (1996). *Understanding Risk: Informing Decisions in a Democratic Society*. Washington, DC: National Academies Press.

Suter, G. W., II. (2008). Ecological risk assessment in the United States Environmental Protection Agency: An historical overview. *Integrated Environmental Assessment and Management*, 4(3), 285–289.

Tai, S. (2010). Comparing approaches towards governing scientific advisory bodies on food safety in the United States and the European Union. *Wisconsin Law Review*, 2010, 627–671.

Thompson, P. B. (2003). Value judgments and risk comparisons: The case of genetically engineered crops. *Plant Physiology*, 132, 10–16.

Vos, E., and Wendler, F. (2006). Food safety regulation at the EU level. In E. Vos and F. Wendler (eds), *Food Safety Regulation in Europe: A Comparative Institutional Analysis* (pp. 65–138). Antwerp and Oxford: Intersentia.

Wickson, F., and Wynne, B. (2012). The anglerfish deception. *EMBO Reports*, 13(2), 100–105.

Winickoff, D., Jasanoff, S., Busch, L., and Grove-White, R. (2005). Adjudicating the GM food wars: Science, risk, and democracy in world trade law. *Yale Journal of International Law*, 30, 81–123.

Wynne, B. (2006). Public engagement as a means of restoring public trust in science: Hitting the notes, but missing the music? *Public Health Genomics*, 9(3), 211–220.

11

Monstrous materialities: ash dieback and plant biosecurity in Britain

Judith Tsouvalis

The aim of the edited volume *Science and the politics of openness* is to raise awareness of the double-sided controversial nature of initiatives aimed at improving relations between science, policymaking, politics and publics. Efforts have been made to strengthen public trust in expert knowledge. These include dialogues organised between scientists and concerned publics on contentious, ethically complex issues, inviting specific publics to help decide the trajectories of controversial scientific and technological innovations and opening up the questions of the role of science in politics and vice versa to closer scrutiny. All this has been much debated in the UK and elsewhere since around the turn of the millennium (House of Lords, 2000; Stilgoe et al., 2006; Wilsdon and Doubleday, 2013). These 'monstrous' sides of relations between science, policymaking, politics and publics – aspects that are unexpected, uncertain, unknown, uncomfortable, preferably ignored and often downplayed– also entail material ones, and these can exert a strong influence over how these relations evolve (Latour, 2004, 2013; Raman and Tutton, 2010; Tsouvalis, 2016; Tsouvalis and Waterton, 2015). This chapter begins with a discussion of the monstrous materiality of Chalara ash dieback (Chalara), a deadly fungal tree disease that has decimated the ash population across Europe since the early 1990s.

On 7 March 2012, Chalara was officially declared present in England, following the routine inspection of a nursery in Buckinghamshire. The ash saplings infected with the disease were found in a consignment of plants imported from the Netherlands. Plant disease outbreaks are on the increase worldwide and many are linked to international trade.

Countless pathogens, insects and animals circulate through the global trade network as travel companions in plants, soil, logs, packaging materials, nursery stock, fruit and seeds (Brasier, 2008: 793–794, 796–797). They pay no heed to political or geographical boundaries, and with changing climate conditions their border crossings are increasingly common and successful. Unfortunately for native, locally adapted plant communities, this is bad news. Generally suffering few ill effects from the life forms they have co-evolved with, they often succumb to encounters with new ones. Chalara first broke out in Poland and Latvia in the early 1990s, having arrived there on infected ash saplings imported from East Asia (Drenkhan et al., 2014; Han et al., 2014; Zhao et al., 2012). The disease reached Germany in 2002, Denmark in 2003, Belgium in 2010 and northern France in 2012.

Given this rapid geographical spread west, it is surprising that the British Government did nothing to try to prevent its arrival in Britain. Part of the reason for this was that scientific knowledge about the cause of the disease was scant and that an error had occurred in its taxonomy. The latter played a brief but important role in allowing Chalara slip through the net of legislation then in place to prevent the trade-related spread of infectious diseases in the EU. It allowed the pathogen to spread freely through mainland Europe and eventually take a foothold in Britain (Freer-Smith et al., 2013: 23). This is the monstrous side of Chalara, a disease that remained *terra incognita* in science for many years, and this is its story.

More than a decade after arriving in Eastern Europe, in 2006 *Chalara fraxinea* was named as the pathogen responsible for the disease (Kowalski, 2006). Three years later, however, new research suggested that Chalara was only a stage – the asexual form, or anamorph – in the life cycle of a fungus called *Hymenoscyphus albidus*, known to science since 1851 and indigenous across Europe and the UK. Historically considered a harmless saprophytic ascomycete, *H. albidus*, which thrives on ash leaves and plays an important role in the nutrient cycle, suddenly assumed the sinister role of the ash-tree killer.

In 2010 molecular studies overturned this verdict, showing that the disease was actually caused during the asexual phase of a newly identified fungus, *Hymenoscyphus pseudoalbidus*. Identical to *H. albidus* in appearance, it is distinguishable from it only by molecular analysis (Queloz et al., 2010). This case of mistaken identity had serious consequences for Britain, as it prevented the British Government from acting on the advice the Forestry Commission (FC) had received from the Horticultural Trade Association in 2009: to impose an import ban on all ash and ash-related products. When asked why during a House of Lords' debate on Chalara in 2012, the Government's reply was that it had 'no reason to believe that this [the discovery of Chalara in the Buckinghamshire nursery] was anything other than an isolated incident' (quoted in Downing, 2012: 10). Probing deeper, however, we find that the FC assumed it was 'dealing with a pathogen already present in the UK and this precluded the UK from initiating an emergency response under the EU Plant Health Directive and World Trade Organization phytosanitary rules and using import restrictions as a means of control' (Downing, 2012: 10).

The head of Plant Health at the FC thus responded to the Horticultural Association's letter that 'our hands are tied' (Downing, 2012: 10). The Forest Research branch of the FC could only issue a pest risk alert to the forestry and horticultural sectors to make them aware of the symptoms of ash dieback. It could not, however, request a full pest risk analysis (PRA). A PRA is a protective measure that all EU member states can apply for under Council Directive 2000/29/EC. Its aim is to prevent the introduction into the EU of organisms harmful to plants or plant products and to stop them spreading in the EU. Chalara's monstrous side – a side only molecular analysis could uncover – illustrates the complex linkages that exist between materiality, policy, legal instruments, human knowledge and understanding, and countless other factors in the emergence of relations between science, politics and publics. After it was declared that *H. albidus* was not the cause of Chalara, an import ban of ash and ash-related products came into effect in Britain on 29 October 2012. By that time, of course, the horse had bolted.

Opening up the science of ash dieback

Knowledge about Chalara was scarce and the public response to the disease in England was exceptionally strong and emotional. As a result, the year 2012 saw an unprecedented opening up of the science of Chalara to scientists internationally and concerned publics locally. This response also needs to be situated in the context of years of funding cuts in the area of plant pathology in the UK, which had led to a steep decline in expertise in this field. In a report published in 2009, the Royal Society had urged universities and funding bodies to collaborate in order to revive the teaching of subjects like agronomy, plant physiology, pathology, general botany, soil science, environmental microbiology, weed science and entomology. This was no mean feat, as an audit of plant pathology undergraduate teaching and training commissioned three years later by the British Society for Plant Pathology (2012) revealed. It found that many plant pathology research institutes and industrial research and development departments had been closed; plant pathologists were ageing; retiring higher education institute plant pathologists were rarely replaced; fewer than half the 103 higher education institutes that offered biology, agriculture, horticulture or forestry courses at BSc level still taught plant pathology, and only half of these offered practical classes. The British Society for Plant Pathology wondered whether higher education institutes would be able to retain their capacity to teach plant pathology in five to ten years' time, given that 'new departmental appointments and RAE/REF assessments are driven in part by the Impact Factor (IF) of scientific publications. The highly specialised nature of much plant pathology research means that many publications are of low IF' (British Society for Plant Pathology, 2012: 2). A key recommendation of the Tree Health and Plant Biosecurity Expert Taskforce (THPBET) set up in November 2012 following the ash dieback outbreak was that 'key skills shortages' in this field needed to be urgently addressed. To combat Chalara, desperate measures were therefore in order. For example, in December 2012, the open-source platform OpenAshDieBack (oadb.tsl.ac.uk) was launched. It had been designed by scientists from the John Innes Centre in Norwich, and invited scientists from around the world to share scientific data on Chalara. This unconventional step of rapidly generating and releasing genomic sequence data was premised on the understanding that 'to foster open science and make it possible for experts around the world to access the data and analyse it immediately [would] speed up the process of discovery' (MacLean, 2016).

Another effort to open up and speed up the science of Chalara was the Facebook-based crowdsourcing game *Fraxinus*, developed by the Sainsbury Laboratory in Cambridge. *Fraxinus* presents players with real reference DNA sequences from the ash tree genome and asks them to match up multiple DNA sequence reads from other samples with the aim of identifying regions of the genome that display characteristics such as resistance. These could then be used to breed new, disease-resistant ash tree varieties. *Fraxinus* and OpenAshDieBack both require mass participation, and between August and December 2013, 51,057 people played the *Fraxinus* game.

Opening up in the context of biosecurity, the conceptual framework within which plant health risks are currently approached, is closely connected to activities like surveillance, monitoring and control, and here citizen science came to play a particularly significant role. The Living Ash Project, funded by the Department for Environment, Food and Rural Affairs (Defra), is one example of an initiative where the public can get involved in monitoring Chalara. Another is the smartphone app AshTag, developed by the Adapt Low Carbon Group at the University of East Anglia and launched in October 2012. Initially, AshTag enabled concerned members of the public to record infected trees and submit photos of them to experts for assessment. These data were then used to map the spread of the disease across the UK. Since 2016 AshTag has been collecting data on healthy trees in the hope of identifying disease-resistant ones. Citizen scientists also help monitor tree diseases through the Open Air Laboratories network. All three initiatives are focused on monitoring, surveillance and, ultimately, disease control. Did these novel and exciting ways of opening up the monstrous materiality of Chalara in science correspond to an equal opening up of the Pandora's box of free trade and its role in the perpetuation of plant disease epidemics to broader political scrutiny and public debate? Did it correspond to an opening up of policymaking in this field?

'Biosecurity': turning a complex socio-political problem into a techno-scientific challenge

This section attempts to answer these questions. It draws on findings from an in-depth qualitative study (Denzin and Lincoln, 2005; Silverman, 2005) conducted in 2014 of the Government's response to ash dieback. The study was funded by the Leverhulme Trust under its Making Science Public programme. Sixteen in-depth, semi-structured interviews were carried out, nine with members of the THPBET and seven with civil servants and experts otherwise involved in supporting and advising the Government on plant biosecurity. The interview schedule covered a broad range of topics, including questions about the interface between science, politics, policymaking and the public, and about how the THPBET worked, made its recommendations, and addressed and resolved conflicting views. Data on these and related issues were also collected from secondary data sources, including newspapers, government documents, non-governmental organisation (NGO) reports, legal documents, social media, TV documentaries and academic literature.

As indicated earlier, when ash dieback was first discovered the British Government had no contingency plan in place to deal with plant disease epidemics. Scientists knew very little about the disease and plant diseases rarely made it into the headlines of the national press. Although the Government had been reminded by the Foresight Project on Infectious Disease in 2006 that 'diseases in plants and animals act as barriers to economic development and also threaten ecosystems' (Foresight, 2006: iv), and urged by the Independent Panel on Forestry in 2012 to 'speed up delivery of the Tree Health and Plant Biosecurity Action Plan by additional investment in research on tree and woodland diseases, resilience and biosecurity controls' (Independent Panel on Forestry, 2012: 34), little had as yet been done to enact these recommendations. At the European level, in response to the steep rise in tree and plant diseases in the EU, the European Commission (EC) had commissioned an evaluation of the EU's plant health regime in 2009.

The key instrument of this quarantine legislative system dating from 1977 is Council Directive 2000/29/EC. It is meant to guard against all pests and diseases, but in practice only targets the most dangerous ones, of which 250 are listed in its annexes. The plant health regime encompasses measures like plant inspections at production sites, during the growing season and post-harvest; producer registration; and issuing plant passports. It forms part of international regulatory frameworks, including the International Plant Protection Convention of the Food and Agricultural Organization of the United Nations, and the World Trade Organization Sanitary and Phytosanitary (Plant Health) Agreement. Their prime objective is to foster free trade: in 'essence, *Biosecurity* balances enthusiasm for international trade with the need to protect against risks' (Manzella and Vapnek 2007: vii; emphasis in the original).

The EC considers the EU plant health regime as 'indispensable for protecting the health, economy and competitiveness of the EU plant production sector as well as for maintaining the Union's open trade policy' (EC, 2013). It describes it as 'unique in that it is an open regime: movements of plants and plant products into and within the Union are allowed' (EC, 2013:1). The 2009–2010 review, however, found this unique regime was thoroughly inadequate in preventing plant disease epidemics and advised that it be modernised through more focus on prevention, better risk targeting (prioritisation) and more solidarity (EC, 2010). In 2013 the EC warned that 'the existing regulatory framework is … unable to stop the increased influx of dangerous new pests caused by the globalization of trade', and predicted that 'high volumes of imports from other continents … imply a high probability of future outbreaks of foreign pests' (EC, 2013: 1). Only a modernised regime, it concluded, could 'effectively address the plant health impacts of globalisation [and] mitigate the plant health impacts of climate change'. Proposals for improving the EU plant health regulations have since been made and are currently under discussion by the European Parliament and Council. It is doubtful, however, that they will bring about greater plant biosecurity as long as the plant health regime remains tied to the objective of fostering free trade, which is considered by some as the greatest threat to plant health (Brasier, 2005, 2008). On the contrary, plant disease epidemics are likely to increase in number.

At this point, we need to take a closer look at the framing of tree and plant pests and diseases as a 'biosecurity' risk. In Britain, the term 'biosecurity' first entered politics in a House of Commons debate on the foot and mouth disease outbreak in 2001. It is thought that because concerns over affairs of state and national security loom large at this level, biosecurity discourse became littered with references to 'border controls' and 'surveillance' (Donaldson, 2008: 1552), and as a result the protection of the 'native' from the 'non-native', 'alien', and 'invasive' (Nerlich et al., 2009). Studies of the effects of the discourse of biosecurity have found it to be highly restrictive, preventing alternative definitions and understandings of disease epidemics from emerging (Hinchliffe and Ward, 2014; Vogel, 2008). They have also found that tensions between biosecurity governance and neo-liberal international trade priorities remain ill understood (Meyerson and Reaser, 2012), that the dominant biosecurity metaphor of security and the fears that underpin it direct resource allocation towards the fortification of boundaries (Nerlich et al., 2009), and that in some countries biosecurity politics are in the process of engineering a new kind of social identity: 'biosecure citizenship' (Barker, 2010).

From a theoretical standpoint, biosecurity discourse can be understood as forming part of the broader trend in Western societies of being risk-averse and overanxious about health, safety and security. Beck's (1992) *Risk Society* thesis, Foucault's (2004, 2007) biopolitics and Latour's (2003) version of Beck's thesis using actor-network theory have all served here as explanatory sources. The conclusion drawn by Defra from the final report of the THPBET exemplifies some of these arguments. In prose littered with military metaphors, Defra urges the UK to be 'better prepared in understanding the risks of what pests and diseases are likely to arrive, when, where and how they might invade, how severe the impact is likely to be and what options are available for interception, eradication, mitigation or adaptation' (Defra, 2013: 2).

Framing plant diseases in this way has far-reaching consequences for policy and democracy. As Duckett et al. (2015) have shown, risk-based policy is based on a positivist epistemology that favours objective, scientific and technical risk assessment rather than an opening up of complex issues to public and political scrutiny and debate. This can lead to a form of post-politics which is exacerbated by consensual policymaking 'in which the stakeholders ... are known in advance and where disruption or dissent is reduced to debates over the institutional modalities of governing, the accountancy calculus of risk, and the technologies of expert administration or management' (Swyngedouw, 2011: 268). While risk-based approaches can constitute a valuable source of knowledge alongside other knowledges and approaches, they cannot, on their own, solve the monstrous aspects of the increasingly tricky and complex problems we face (Chilvers and Kearnes, 2016; Grove-White et al., 2006). This was the conclusion drawn by members of the THPBET on the process in which they were involved in addressing Chalara.

Public concern over Chalara was great. Apocalyptic imaginaries of a landscape devoid of ash trees and bleak economic forecasts of the consequences of the disease flooded the pages of national newspapers and social media sites. This response took Defra by surprise, and it duly commissioned a study of the THPBET by a social scientist to better understand it (Pidgeon and Barnett, 2013). The study concluded that Defra was dealing with a case of the 'social amplification of risk', where numerous, often lingering, anxieties culminate to find expression in response to a particular event. The Government reacted to to this response by convening a national emergency (Cabinet Office Briefing Rooms – COBR) meeting in London in November 2012, with the aim of showing people 'how seriously the Government is taking the threat of this disease' (Defra, 2012a). It also commissioned the FC to carry out a rapid, large-scale survey to establish the extent and spread of the disease, as the National Forest Inventory of 2009–2012 had recorded 103 diseased ash trees among the 15,000 inspected, none of which were infected with Chalara (House of Lords, 2012: 10). Finally, it set up the THPBET.

This taskforce was entirely composed of 'Chief Scientific Advisors and eminent Government and academic experts' (Beddington, quoted in House of Commons Library, 2012: 1). Its remit was to comment and advise on Defra's scientific evidence and approach to Chalara and 'the current threats from pests and pathogens'. It was also tasked with making 'recommendations about how those threats to trees could be addressed' (Defra, 2012a: 7). The names of the taskforce participants are listed in its Final Report (Defra, 2013: 49). Of its fourteen members, eleven held professorships at the time and all fourteen were educated to PhD level. Ten were natural scientists, four were social scientists, and of these, two were economists. The taskforce was supported by a public sector officials advisory group, whose members were drawn from Defra and the Defra network organisations. External referees were invited to comment on the reports, as were a broad range of stakeholders. The terms of reference for the taskforce were determined before it first convened. The language in which they are formulated is indicative of the risk-based approach adopted, containing references to 'best available evidence', an 'assessment of risk status', 'appropriate risk assessment tools', a 'rapid evidence assessment', a 'risk mitigation framework', 'contingency planning', and 'emergency response arrangements'. Stakeholders and the public played no role in framing the problem of ash dieback at this stage, nor did they have a say in who ought to address a challenge of such magnitude. Many stakeholders were concerned about this, as the empirical study conducted by the author and described at the beginning of this section found:

> Conservation organisations in particular were quite critical that it [the taskforce] was set up without any sort of conversation with them about membership. ... They would have liked to have had an opportunity to have suggested how the Terms of Reference [were] framed. ... They were ... invited [to] sit on the Stakeholder Advisory Panel ... after the Terms of Reference and membership had been made. (Respondent 11)

The deadlines for the publication of the two reports to be produced were determined from the outset. The interim report was due by the end of November 2012 (two weeks after the first meeting of the taskforce) and a final report by the spring of 2013 (Defra, 2012b). These tight deadlines – indicative of the perceived emergency of Chalara – greatly impacted on the speed with which the taskforce had to work. The taskforce itself was set up within days of the COBR meeting and was purposely kept small. The names of most of its members were proposed by Defra. Those able to participate at such short notice had to be available for meetings immediately and be committed to working to tight deadlines. Apart from several two-day meetings, telephone conferences and email exchanges took place and participants were assigned to expert groups that tackled specific issues. They had to review and comment on vast numbers of documents. Senior plant health officials were actively involved in meetings and the chief scientific advisor, the chairman of the taskforce, the Secretary of State and Minister Lord De Mauley met on a regular basis to discuss any progress made.

This tight timeframe and the predefined terms of reference of the THPBET, together with the fact that the taskforce was primarily composed of experts with existing links to the government department they now advised, are characteristic of the technocratic post-political approach to risk-based problem solving and policymaking described by Duckett et al. (2015) and Swyngedouw (2011) above. Their combined effects were explored during interviews. The findings suggest that they impacted negatively both on the degree of stakeholder and public involvement and on the degree to which disagreement and conflict could emerge and be addressed during the THPBET meetings.

Concerning stakeholder and public participation, one interviewee observed that the taskforce was not a public forum in the sense that there were no public meetings and no public dialogue took place. Although 'additional people [from local authorities, trade associations, environmental interest groups and others] who were not members of the taskforce ... were brought into meetings and provided written evidence' (respondent 5), they were invited only to comment on materials already produced by the taskforce:

> There were several meetings convened, each one had a very specific agenda that was marginally directed towards arriving at a useful set of recommendations that could be justified on the basis of the scientific background. ... There was also a later phase where ... there was an attempt – prior to submission of these recommendations – to basically get input from various UK Stakeholder groups. (Respondent 8)

Following the first two-day meeting of the taskforce in November 2012, an interim report containing eight recommendations was published (Defra, 2012). Reviewers for the report were chosen on the basis of their ability 'to constructively contribute to the objectives of the taskforce', and a broad range of stakeholders were invited to comment on it (respondent 6). Significantly, this respondent points out that the 'Interim Report came out first and then they [the taskforce] used that to refine what they thought their recommendations should be and they talked to stakeholders'. The stakeholders could therefore only refine conclusions already drawn by the experts and formulated as recommendations in their first report. Even then, there was little room to accommodate their views:

> We were trying to make our recommendations based on science. So we weren't really trying to make them fit with the views of stakeholders at all. ... There was much less stakeholder input into the expert report because it was not meant to be an exercise which drew its information from stakeholders. It was meant to be an exercise that drew its information from ... the best understandings of ... both natural science and social science. (Respondent 9)

Time, bureaucratic procedures, scientific knowledge, and the Government's framing of tree and plant health as a problem of biosecurity all impacted on the issues the taskforce could address and the recommendations it finally made. Taskforce meetings generally lasted under an hour. This made 'a deeper engagement and the development of conversations' impossible (respondent 15). However, it proved a powerful strategy for keeping conflict at bay: 'there is a risk ... that if views were polarised they could be very polarised by the end of three hours' (respondent 15).

Unsurprisingly, most interviewees reiterated the view that there had been little disagreement and conflict between taskforce members during meetings. Some respondents, however, were unhappy about this and thought that 'there had been things that had not been included in the reports' and that 'other recommendations could have been made' (respondent 11). This respondent felt that general agreement was at least in 'part to do with the way the discussions were framed' and 'there was certainly the impression that controversial issues were avoided'.

The taskforce's recommendations reflect this. They include the development of a prioritised UK plant health risk register; the appointment of a chief plant health officer responsible for the UK plant health risk register and for providing strategic and tactical leadership for managing risks; the development and implementation of procedures for preparedness and contingency planning to predict, monitor and control the spread of pests and pathogens; and the revision, simplification and strengthening of governance and legislation. They also include the recommendation that epidemiological intelligence from the EU and other regions needs to be better used, and EU regulations for tree health and plant biosecurity improved. Biosecurity at the border and in the UK needs to be strengthened, capabilities and communications improved through the development of a modern, user-friendly system providing quick and intelligent access to information about tree health and plant biosecurity and key skills shortages addressed (Defra, 2013: 5).

Many taskforce members described these risk-orientated recommendations as limiting. They were, one observed, 'quite technical and cathedral and as a result less controversial'. It had been easy, the respondent explained, to reach agreement on the need for the appointment of a new chief plant health officer and the creation of a risk register (respondent 11). Another saw them as 'rather sort of bureaucratic-type recommendations' (respondent 2). Some taskforce members had expressed concern during meetings about the 'risk-orientated approach' to biosecurity, saying they would have preferred a 'pathways approach'.

However, they knew that such a paradigm change would have proved controversial with the nursery trade, as it would have opened up room for a critical appraisal of the role of trade and the single market in the spread of plant disease, the checking of plants for disease prior to them being moved, consumer behaviour, and the biosecurity implications of the work of professionals such as landscape architects. Unfortunately, debating such issues was beyond the remit of the THPBET. It could therefore not address key drivers of plant disease epidemics, including the effects of trade, even though taskforce members 'were all of the view that it would be much better if the UK could impose trade restrictions for plant health reasons' (respondent 4). Indeed, the respondent went on, 'We [the taskforce] ought to say, if you really want to tackle this you need to ban import on plants, which would be politically not useful at all. … we found it more difficult to see impossibilities in the human world than in the natural world. … and where you see impossibilities affects how you make recommendations.' Echoing this view, another respondent lamented that 'you can make as many recommendations as you like, but the science can't sort those issues out' (respondent 14).

Conclusion: the many monsters of plant biosecurity

The risk-based approach to ash dieback adopted by the Government in 2012 in response to Chalara transformed a highly contentious socio-economic, political and material problem of monstrous and messy proportions into a neatly defined techno-scientific challenge. As a result, trade-related plant disease outbreaks continue to be an issue where 'debate is not only seriously lacking but may also be suppressed through non-recognition or even avoidance of the issues' (Brasier, 2005: 54, 2008; Daszak et al., 2000). Such issues, as observed by the THPBET member interviewed above, cannot be sorted out by science.

Although ash dieback catapulted plant health to the top of the Government's agenda in 2012, many of its monsters remain lurking in the dark. The science of plant pathology was opened up in novel and exciting ways both to scientists and to the public after the strong public reaction to Chalara, which led to the long-overdue allocation of resources in this area. However, interpreting these efforts as a democratisation of science or as 'making science public' would be a mistake. Rather, they formed part of the dominant risk-based approach to plant biosecurity endorsed by the Government and were primarily directed at surveillance, monitoring and plant disease control and at the changing of the very nature (the genetic makeup) of the life forms affected by Chalara, ash trees. The risk-based approach adopted also meant that the THPBET was structured and designed in ways that make it a perfect example of consensual post-political policymaking.

A deeper engagement with the complex economic, socio-cultural, material and political drivers behind tree and plant disease epidemics was impossible. This raises serious questions about the role scientists, social scientists and humanities scholars are often made to play in policymaking, especially in the case of emergencies. Concerning plant health, for example, there are plenty of studies that document the detrimental effects of trade and certain horticultural practices, such as the overuse of herbicides and pesticides in nurseries, the importation of live trees, or the practice of exporting seeds and importing saplings to save labour costs, on plant health. If their findings were more powerfully articulated and taken seriously by government institutions, this would inevitably put the spotlight on politically more delicate and challenging issues, and it is these issues that urgently need addressing in this field. To simply exploit scientific 'evidence' for the purpose of upholding neo-liberal trade arrangements or finding ever more life-transforming technologies to counteract their costs is not only irresponsible, it is also deeply unethical.

References

Barker, K. (2010). Biosecure citizenship: Politicizing symbiotic associations and the construction of biological threat. *Transactions of the Institute of British Geographers, New Series*, 35(3), 350–363.

Beck, U. (1992). *Risk Society: Towards a New Modernity*. London: Sage.

Brasier, C. (2005). Preventing invasive pathogens: Deficiencies in the system. *Plantsman, New Series*, 4, 54–27.

Brasier, C. (2008). The biosecurity threat to the UK and global environment from international trade in plants. *Plant Pathology*, 57, 792–808.

British Society for Plant Pathology. (2012). *Plant Pathology Education and Training in the UK: An Audit*. British Society for Plant Pathology. Retrieved 7 July 2014 from: www.bspp.org.uk/society/docs/bspp-plant-pathology-audit-2012.pdf.

Chilvers, J., and Kearnes, M. (eds) (2016). *Remaking Participation: Science, Environment and Emergent Publics*. London: Routledge.

Daszak, P., Cunningham, A. A., and Hyett, A. D. (2000). Emerging infectious diseases of wildlife: Threats to biodiversity and human health. *Science*, 287, 443–449.

Defra (2012a). *Government Action on Ash Tree Disease Chalara. Gov.uk*, press release, Defra, 2 November. Retrieved 2 June 2015 from: www.gov.uk/government/news/government-action-on-ash-tree-disease-Chalara.

Defra (2012b). *Tree Health and Plant Biosecurity Expert Taskforce: Interim Report*. Ref PB13842. *Gov.uk*, policy paper, Defra, 30 November. Retrieved 2 June 2015 from: www.gov.uk/government/publications/tree-health-and-plant-biosecurity-expert-taskforce-interim-report.

Defra (2013). *Tree Health and Plant Biosecurity Expert Taskforce: Final Report*. Ref PB13878. *Gov.uk*, policy paper, Defra, 20 May. Retrieved 7 July 2014 from: www.gov.uk/government/publications/tree-health-and-plant-biosecurity-expert-taskforce-final-report.

Denzin, N. K., and Lincoln, Y. S. (2005). *Handbook of Qualitative Research*, 3rd edition. Thousand Oaks, CA: Sage.

Donaldson, A. (2008). Biosecurity after the event: Risk politics and animal disease. *Environment and Planning A*, 40, 1552–1567.

Downing, E. (2012). *Ash Dieback Disease: Chalara fraxinea*. Standard note SNSC6498. [London]: House of Commons Library.

Drenkhan, R., Sander, H., and Hanso, M. (2014). Introduction of Mandshurian ash (*Fraxinus mandshurica Rupr.*) to Estonia: Is it related to the current epidemic on European ash (*F. excelsior L.*)? *European Journal of Forest Research*, 133, 769–781.

Duckett, D., Wynne, B., Christley, R. M., Heathwaite, A. L., Mort, M., Austin, Z., Wastling, J. M., et al. (2015). Can policy be risk-based? The cultural theory of risk and the case of livestock disease containment. *Sociologia Ruralis*, 55(4), 379–398.

EC (2010). *EU's Plant Health To Be Strengthened and Better Protected through New Legislation*. Press release IP/10/1189. Retrieved 22 August 2017 from: http://europa.eu/rapid/press-release_IP-10-1189_en.htm.

EC (2013). *Proposal for a Regulation of the European Parliament and of the Council on Protective Measures against Pests of Plants*. COM(2013) 267 final. Brussels: EC. Retrieved 8 April 2014 from: http://ec.europa.eu/dgs/health_food-safety/pressroom/docs/proposal-regulation-pests-plants_en.pdf.

Foresight (2006). *Infectious Diseases: Preparing for the Future*. Executive summary. London: Office of Science and Innovation. Retrieved 14 July 2014 from: www.gov.uk/government/uploads/system/uploads/attachment_data/file/294243/06-760-infectious-diseases-report.pdf.

Foucault, M. (2004). *Society Must Be Defended*. London: Penguin.

Foucault, M. (2007). *Security, Territory, Population: Lectures at the College de France 1977-78*. London: Palgrave Macmillan.

Freer-Smith, P., Ward, M., and Simmonds, M. (2013). Ash trees: Effects of *Chalara fraxinea*. *Science in Parliament*, 70(1), 23-28.

Grove-White, R., Kearnes, M. B., Macnaghten, P. M., and Wynne, B. (2006). Nuclear futures: Assessing public attitudes to new nuclear power. *Political Quarterly*, 77(2), 238-246.

Han, J. G., Shrestha, B., Hosoya, T., Lee, K. H., Sung, G. H., and Shin, H. D. (2014). First report of the ash dieback pathogen *Hymenoscyphus fraxineus* in Korea. *Mycobiology*, 42(4), 391-396.

Hinchliffe, S., and Ward, K. J. (2014). Geographies of folded life: How immunity reframes biosecurity. *Geoforum*, 53, 136-144.

House of Commons Library (2012). *Ash Dieback*. Debate pack. [London]: House of Commons Library. Retrieved 22 July 2015 from: www.sarahnewton.org.uk/files/ashdieback.pdf.

House of Lords (2000). *Select Committee on Science and Technology (2000) Science and Society*. London: The Stationery Office.

House of Lords (2012). Trees: British ash tree. Question for short debate asked by the Earl of Selborne. *Hansard*, 5 November, column 862-882. Retrieved 9 January 2014 from: www.publications.parliament.uk/pa/ld201213/ldhansrd/text/121105-0003.htm#12110554000087.

Independent Panel on Forestry (2012). A woodland culture for the 21st century. In *Independent Panel on Forestry: Final Report* (pp. 14-43). Retrieved 20 September 2013 from: www.gov.uk/government/uploads/system/uploads/attachment_data/file/183095/Independent-Panel-on-Forestry-Final-Report1.pdf.

Kowalski, T. (2006). *Chalara fraxinea* sp. nov. associated with dieback of ash (*Fraxinus excelsior*) in Poland. *Forest Pathology*, 36, 264-270.

Latour, B. (2003). Is re-modernization occurring and if so, how to prove it? A commentary on Ulrich Beck. *Theory, Culture and Society*, 20(2), 35-48.

Latour, B. (2004). *The Politics of Nature: How to Bring the Sciences into Democracy*. Cambridge MA: Harvard University Press.

Latour, B. (2013). *An Inquiry into Modes of Existence*. Cambridge MA: Harvard University Press.

MacLean, D. (2016). *Team MacLean – Bioinformatics @ The Sainsbury Laboratory*. Accessed 9 December 2016 at: www.danmaclean.info/research.html.

Manzella, D., and Vapnek, J. (2007). *FAO Legislative Study 96: Development of an Analytical Tool to Assess Biosecurity Legislation*. Rome: Food and Agricultural Organization of the United Nations.

Meyerson, L. A., and Reaser, J. K. (2012). Biosecurity: Moving towards a comprehensive approach. *BioScience*, 52(7), 593–600.

Nerlich, B., Brown, B., and Wright, N. (2009). The ins and outs of biosecurity: Bird flu in East Anglia and the spatial representation of risk. *Sociologica Ruralis*, 49(40), 344–359.

Pidgeon, N., and Barnett, J. (2013). *Chalara and the Social Amplification of Risk*. Ref. PB 13909. *Gov.uk*, Defra. Retrieved 25 November 2016 from: www.gov.uk/government/policy-advisory-groups/tree-health-and-plant-biosecurity-expert-taskforce.

Queloz, V., Gruenig, C. R., Berndt, R., Kowalski, T., Sieber, T. N., and Holdenrieder, O. (2010). Cryptic speciation in *Hymenoscyphus albidus*. *Forest Pathology*, 4, 133–142.

Raman, S., and Tutton, R. (2010). Life, science and biopower. *Science, Technology, and Human Values*, 35(5), 711–734.

Silverman, D. (2005). *Doing Qualitative Research: A Practical Handbook*, 2nd edition. London: Sage.

Stilgoe, J., Irwin, A., and Jones, K. (2006). *The Received Wisdom: Opening Up Expert Advice*. London: Demos.

Swyngedouw, E. (2011). Depoliticized environments: The end of nature, climate change and the post-political condition. *Royal Institute of Philosophy Supplement*, 69, 253–274.

Tsouvalis, J. (2016). Latour's object-orientated politics for a post-political age. *Global Discourse*, 6(1–2), 26–39.

Tsouvalis, J., and Waterton, C. (2015). On the political nature of *Cyanobacteria*: Intra-active collective politics in Loweswater, the English Lake District. *Environment and Planning D: Society and Space*, 33(3), 477–493.

Vogel, K. M. (2008). Framing biosecurity: An alternative to the biotech revolution model? *Science and Public Policy*, 35(1), 45–54.

Wilsdon, J., and Doubleday, R. (2013). Hail to the chief: Future directions for scientific advice. In R. Doubleday and J. Wilsdon (eds), *Future Directions for Scientific Advice in Whitehall* (pp. 7–20). Centre for Science and Policy, University of Cambridge; Science Policy Research Unit and ESRC STEPS Centre, University of Sussex; Alliance for Useful Evidence; Institute for Government; and Sciencewise. Retrieved 25 June 2013 from: www.csap.cam.ac.uk/media/uploads/files/1/fdsaw.pdf.

Zhao, Y. J., Hosoya, T., Baral, H. O., Hosaka, K., and Kakishima, M. (2012). *Hymenoscyphus pseudoalbidus*, the correct name for *Lambertella albida* reported from Japan. *Mycotaxon*, 122, 25–41.

12

An Inconvenient Truth: a social representation of scientific expertise

Warren Pearce, Brigitte Nerlich

On 30 June 2006 *An Inconvenient Truth* (*AIT*) (Guggenheim, 2006), a climate-change documentary presented and written by leading US Democrat politician Al Gore, was released. The film contains a heady mix of expert scientific evidence, personal stories and normative political statements. An 'oral history', based on interviews with those involved in the creation of the film and celebrating this anniversary, proclaimed: 'Somehow, a film starring a failed presidential candidate and his traveling slideshow triggered a seismic shift in public understanding of climate change' (Armstrong et al., 2016).

It is likely that *AIT* has contributed as much as anything or anyone to making climate-change expertise public. In particular, it brought climate-science expertise, which had steadily accumulated in the preceding decades, into the public realm in a new way: combining scientific data with personal stories and calls for political action. In combining these elements, *AIT* made climate change public by offering a particular social representation of climate change. While primarily appealing to a public that was already interested in, and attentive to, climate change, *AIT* also helped to broaden that audience. The film's intended audience was what one may call its 'convenient' public. On the other hand, the film's very success in speaking to such a public also triggered contestation from what one may call an 'inconvenient' public; that is, from an audience that disputed the film's social representation of climate-change expertise – in some cases the film and/or its producer were framed as 'monstrous'. The film thus became a successful meme and what some saw as a dangerous monster at the same time.

In this chapter we discuss *AIT* as an example of taking climate-change expertise out of the pages of science journals and into the public sphere. We draw on the ideas of John Dewey (1938, 1989) and their elucidation by Mark Brown (2009, and see chapter 9) to show how the notion of expertise is the key to understanding the film's motivation, successes and critics. While the purpose of the documentary was to persuade its audience of the consensual truth imparted by climate-science experts, its effect was to become a lightning rod for disagreeing with, criticising and debating with that expertise. Overall, *AIT* created a dominant representation of climate change, based on expertise that became a touchstone for consent and dissent, action and reaction. This position was enhanced by the joint award of the 2007 Nobel Peace prize to Gore and the Intergovernmental Panel on Climate Change (IPCC).

In the following we shall first provide some background to the film's emergence, highlighting its echoes of Dewey's argument that expert knowledge should be integrated in society (Brown, 2009: 150). We use the concept of social representation (Moscovici, 1988) to show how Gore combined scientific content with a personal and political context in order to provide a meaningful representation of climate-change expertise. We highlight how *AIT* sought to create its own public for scientific expertise, returning climate-science expertise to society as one of the many tools with which citizens make sense of the world and solve problems (Brown, 2009: 160–161). We then show how the very elements that helped *AIT* to establish a dominant social representation of climate change also contributed to the creation of a counter-representation and counterpublic that questioned how *AIT* represented climate-science expertise. With *AIT*'s success in bringing social context to scientific content came inevitable contestation. We conclude with some tentative lessons for science communicators from the *AIT* story.

Background

AIT had a huge cultural and political impact following its release in 2006, winning a host of awards, including the 2007 Academy Award for Best Documentary (IMDb, 2015), helping Gore win a share of a Nobel Peace prize with the IPCC and providing an anchor for intense, prolonged debates about climate change.

The documentary was timely, which helped it to embed itself in global culture and shape both dominant or hegemonic and counter-hegemonic polemical social representations of climate change. A dominant or hegemonic social representation is one that is a coercive and widely shared construction of climate change, while a polemic one is defined as 'one which is generated in the course of social conflict, and characterised by antagonistic relations between groups' (Jaspal et al., 2014). In 2007 the IPCC released its *Fourth Assessment Report*, which marked a step change in the public visibility of climate science. These events represented a political and cultural reinforcement of the emerging scientific consensus and were significant in establishing for the first time a dominant, hegemonic representation of climate change that called for significant personal and political action to address the challenge. *AIT* did not disappear from cultural consciousness after 2007. Gore made sure that future campaigns such as Climate Reality built on its success, seeking to train volunteers as 'Climate Reality Leaders ... spreading the word about *the truth* of climate change and the solutions we have today in over 100 countries, making a global challenge a personal issue for citizens on every continent' (Climate Reality, n.d.; emphasis added).

This suggests that *AIT* was a highly successful project, both as a cultural event in itself and as a way of bringing meaning to climate change and momentum to climate-change mitigation. *AIT*'s combination of scientific ideas with personal stories and political activism echoes Dewey's call for 'bare ideas' to have 'imaginative content and emotional appeal' in order to be effective (Dewey, 1989: 115). *AIT* also takes seriously Dewey's notion that scientific expertise is a social product rather than the result of individual scientific brilliance and that science communication marks the return of knowledge to its rightful owners: the public (Brown, 2009: 150). Indeed, *AIT* takes this one step further by seeking to empower its audience to gain the expertise to go out and disseminate locally. Yet, while Dewey points to the seeds of *AIT*'s success, he also shows how the successful communication of scientific knowledge and its social consequences brings more public scrutiny to bear on expertise (Brown, 2009: 159).

A decade later, *AIT* remains an important representation of climate-change expertise. Gore's name continues to be synonymous with public discussions of climate change (Grundmann and Scott, 2014) and *AIT* continues to act as a salient reference point for climate-change critics (e.g. Booker, 2015; *Daily Mail Comment*, 2015; Turnbull, 2011). In the next section we describe the key elements of this representation.

Representing climate-change expertise

Climate science is an example of the scientific representation of nature that responds to a problematic situation. Communicating this expert knowledge is important as the problematic situation is bound up with social conditions (Brown, 2009: 160). Yet Dewey understands that if this expert knowledge is to gain purchase within societies, it must be communicated aesthetically and imaginatively (Brown, 2009: 150). As discussed above, this provides a rationale for *AIT* but it also shows that *AIT* is a social representation of a scientific representation of nature (namely, the abstract concept of climate change). Hence, concepts from social-representations theory help to show how *AIT* represented climate-science expertise by objectifying climate change through humans (personification) and non-humans (ontologisation) (Jaspal et al., 2014). This constituted an attempt to establish a coherent, hegemonic social representation of climate-science expertise that would gain purchase with the *AIT* audience, inspiring them to take various actions on climate change or to contest such actions (Hollin and Pearce, 2015; Jacobsen, 2011; Jaspal et al., 2014; Nolan, 2010).

According to social-representations theory, a social representation is 'a system of values, ideas and practices' about a given social object (Moscovici, 1973: xiii), as well as 'the elaborating of a social object by the community for the purpose of behaving and communicating' (Moscovici, 1963: 251). Such a representation provides a social group with a shared social reality and common consciousness of a particular social object. The primary function of a social representation is to allow a social group to incorporate 'something unfamiliar and possibly troubling into their own network of categories (Moscovici, 1981: 193). Hegemonic social representations are shared by members of a group; they are coercive and uniform. Polemic representations are generated in the course of social conflict and are characterised by antagonistic relations between groups (Jaspal and Nerlich, 2014: 124–125; Moscovici, 2000: 28).

Objectification is the process whereby unfamiliar and abstract objects are transformed into concrete and objective common-sense realities. Moscovici and Hewstone (1983) postulate three subprocesses associated with objectification; namely, the personification of knowledge, figuration and ontologisation. We focus here on the first and the last. The personification of knowledge links the abstract object to a person or a group, providing the object with a more concrete existence through this association. Ontologisation refers to the process whereby physical characteristics are attributed to a non-physical entity, essentially 'materialising' the immaterial.

We will show that while *AIT* helped to elevate the cultural significance of climate change and contributed to forging and disseminating a hegemonic representation of climate change, it also prompted the emergence of a strengthened polemic-representation counterpublic that placed *AIT*'s representation of climate-science knowledge under intense scrutiny. By highlighting some scientific weaknesses in the film and Gore's role as the face of expertise, the counterpublic sought to establish a counter-hegemonic or polemical social representation of climate change. Here, the monsters lurking under the public face of climate change came to life, most notoriously in an episode of *South Park* where Gore was depicted warning of an implausible, unseen monster called ManBearPig (Parker, 2006; Delingpole, 2010). Monstrous representations continue to this day, with a Breitbart article confusingly describing a new sequel to the film (Cohen and Shenk, 2017) as a 'scientific monstrosity' while referring to climate change as 'a non-science beast' (Williams, 2017).

While scientific knowledge plays an important role in the film, Gore evidently recognised, like Dewey, that public mobilisation requires climate change to be made meaningful, not abstract, by manipulating both cognitions and emotions (Beattie et al., 2011) so that 'enough people lock into the same narrative and connect the dots and feel the danger facing their children' (Bates and Goodell, 2007). The emergence of scientific knowledge about climate change has given rise to 'an impersonal, apolitical, and universal imaginary of climate change' that has taken over from 'normative imaginations of human actors engaging directly with nature' (Jasanoff, 2010: 235). *AIT* attempts to redress this balance by personalising and ontologising climate change. Most obviously, it positions Gore – for better or worse – as the human face of the climate-change debate (Jaspal et al., 2014: 114). Yet it also contains other attempts at personalisation. In a powerful early section, Gore tells how his young son was almost killed in a car accident, and of the painful days spent at his bedside waiting to see if he would recover. The parallel is drawn between Gore's son and the natural world that we assume to be stable, showing that the things that we take the most for granted can be taken away from us unexpectedly (Murray and Heumann, 2007).

As well as this personalisation of climate or nature, the film seeks to reintroduce the personal into the accumulation of scientific knowledge. Knowledge is given credence not only using charts and numbers, but by the scientists who produced them. Gore refers to palaeoclimatologist Lonnie Thompson as 'my friend' when arguing that Thompson's research shows a striking correlation between atmospheric carbon dioxide concentrations and temperature. Science may achieve its heft through abstraction (Jasanoff, 2010: 234), but Gore reminds his audience that scientific practice is irreducibly human, through his account of his son's accident.

AIT also seeks to mitigate abstraction through the ontologisation of climate change by way of various non-human forms. The film begins with a paean to the central role of nature in Gore's early life, which is subsequently referenced in the story of his son's car accident. This 'environmental nostalgia' makes climate change real by presenting it as an emotional threat to our own memories of living in nature (Murray and Heumann, 2007). Glaciers are used as another material example of what we might lose from climate change. However, this was not without controversy. One supportive climate scientist's review of *AIT* argued that while the general point was well made, the particular examples used in the film were poorly chosen, as they were probably unrelated to temperature change (Steig, 2008). *AIT* ties climate change to the threat of extreme weather, traumatically felt in the USA through Hurricane *Katrina* just prior to the film's release (Nerlich and Jaspal, 2014). While *Katrina* is mentioned prominently in the film, the important role of engineering failures in the devastation it caused are overlooked; a position described by Rayner as 'using bad arguments for good causes' (2006: 6).

Criticisms of some of the specific examples used in *AIT* highlight a broader tension underpinning the ontologisation of climate change;

that is, that local examples of climate-change-related events are likely to be less scientifically certain than global representations of climate (Hollin and Pearce, 2015). This is not to say that *AIT* is entirely unsuccessful on this front; merely that scientific representations and social representations may often come into conflict. Evaluating how these are resolved depends on whether Gore's role in *AIT* is 'as a politician, a lay expert, or a spokesperson for science' (Hulme, 2009: 81), something that remains unclear during the film.

This section has outlined the social representation of climate-science expertise in *AIT*. The next section demonstrates the integral role of the audience in this representation, as Gore returns science to the people (Brown, 2009: 160).

Emergence of a public

Empire magazine's five-star review of *AIT* begins with an inauspicious synopsis: 'On the face of it, this is the least appealing film in history. A failed politico … preaching to the world about global warming with the aid of PowerPoint' (O'Hara, 2015).

Presentation software such as PowerPoint or Keynote[1] appears to be a questionable medium through which to persuade an audience of the seriousness of climate change. Even at the time of *AIT*'s release, such software was becoming notorious for homogeneous, ready-made slide designs resulting in boring corporate presentations (Reynolds, 2005; Tufte, 2003). While Gore's professionally designed slides avoid the template trap, one might wonder why he chose to make such a presentation the focus of the film, rather than the front line of climate change where the physical effects are beginning to be noticed, as subsequent films have done (Orlowski, 2012). In short, *AIT* foregrounded the presentation as that was the tool with which Gore's message would be propagated by his helpers, supporters and acolytes.

Gore makes clear his frustration with inaction on climate policy from the US Congress and the then Bush administration, using this as the basis for a 'bottom-up' approach to spreading his message 'city by city, street by street, house by house'. Gore explains that he has been 'trying to tell this story for a long time' and that he is focused

---

1 Gore's presentation was developed using Keynote (Reynolds, 2007).

on 'getting people to understand' climate change. Clearly, this is not public education as a good in itself; the intermingling of the positive and the normative points towards the need for the climate-change challenge to prompt particular actions.

AIT ends on an upbeat note, claiming that we already have the technologies available to switch from fossil fuels, and that all that is stopping us is a shortage of political will. The film ends by fading to black, as the text 'Are you ready to change the way you live your life?' appears on the screen, followed by an intermingling of the film's credits with a mixture of tips on reducing personal environmental impacts (e.g. switch to a hybrid car) and bringing about political change (e.g. ask your senators what they are doing about climate change). Viewers were also directed to a supporting website including more details about the film and about climate science, and suggested actions for the audience to undertake ('*An Inconvenient Truth* > take action', 2006). Taken together, the film, website and accompanying book (Gore, 2006) represented a multimedia take on a very traditional linear model of science education, with the idea that presenting members of the public with more scientific information will prompt them to take action. While this is a clear aim of *AIT*, the film also operated at a more sophisticated metalevel.

Gore is the film's sole cast member, but his audience – his intended public – plays an important supporting role throughout. The first faces to appear in the film are those of the attendees at the various presentations of Gore's slideshow around the USA. *AIT*'s main presentation is staged in a way that ensures the audience's faces are often in view, brightly lit and seated in a horseshoe formation. These are not just the faces of people listening to Gore's story, but of those who may retell it to their peers. Soon after the film's release, Gore led a programme of training for people who wanted 'to tell their friends, families and neighbours that human activities are altering global climate and that each person can do something about it' (Haag, 2007). The programme continues today through the Climate Reality Leadership Corps that encourages peer-to-peer communication and 'spreading the word about the truth of climate change' (Climate Reality, n.d.).

In this way, *AIT* went beyond public education to instead aim explicitly at the creation of a climate-change public. For a while Al Gore became known as the high priest spreading an 'environmental gospel' (Mr Americana, 2015; Nerlich and Koteyko, 2009), a title that also contributed to conjuring up the counterpublic that the film did not intend to create. Overall, then, it was not just the content of the slideshow that was important, it was also the performance of the slideshow that is a central part of the film. The film was intended not only to persuade but to have a much stronger performative force: to create a public that in turn would continue the performance. In Dewey's terms, scientific expertise is reinstated as 'a refinement of commonsense inquiry' rather than 'a foreign way of knowing to be imposed on the common sense of an ignorant public' (Brown, 2009: 160). However, this overt focus on putting scientific expertise back into the hands of society was turned back on *AIT* itself, as a counterpublic questioned the film's representation of climate-science expertise.

Emergence of a counterpublic

The evidence presented thus far suggests that *AIT* was extremely successful, not just as a film in its own right but also in establishing a powerful social representation of climate change, an idea that had been somewhat nebulous up to that point. *AIT* was also successful in creating a public actively engaged in reproducing the representation of climate change by training individuals to give presentations based on *AIT* locally. However, individuals are not merely passive recipients of representations; they actively contribute to the construction of new representations in response (Jaspal et al., 2014: 116). Some of these individuals assumed a much more critical view of *AIT* and Gore.

Scepticism about climate science predated the film's release as an important part of the 'struggles over meaning and values in US climate science and politics' (Lahsen, 2008: 216). While such struggles were continuing, US climate politics pre-*AIT* was broadly characterised by a lack of federal-level progress on legislation to cut greenhouse gases. Congress's comprehensive rejection of the Kyoto Protocol was followed by Gore's loss to George W. Bush in the 2000 presidential election, with the subsequent Bush presidency being noted for a stalemate on climate policy. The success of *AIT* towards the end of the Bush presidency provided a window for reframing the US climate debate (Fletcher, 2009: 807). It also acted as a powerful rallying point for climate critics, both in the mainstream media and the blogosphere, who were opposed to more stringent action on greenhouse gases.

A struggle ensued over the film's accuracy, and as *AIT* gained greater public visibility a counterpublic emerged that sought to destabilise the apparently coherent meaning of climate change provided by *AIT* and Gore's newfound position as a public expert. This counterpublic was mobilised through the emerging new media of blogs such as *Watts Up With That* (Watts, 2006) and *Climate Audit* (McIntyre, 2006), as well as syndicated columns in the mainstream media (Elsasser and Dunlap, 2013). The movement challenged the links claimed between climate change and material events (Hulme, 2010), and the credibility of Gore himself (Elsasser and Dunlap, 2013).

It is unsurprising that Gore, as a prominent Democratic politician, became a focus of much conservative commentary. A study of conservative op-eds found him to be by the far most discussed topic related to climate change (Elsasser and Dunlap, 2013: 763). Within the sceptical blogosphere, the three blogs found by Sharman (2014) to be the most central – *Watts Up With That*, *Jo Nova* and *Climate Audit* – have all had numerous posts on Al Gore and/or *AIT*. While Sharman notes that these blogs are more likely than mainstream media op-eds to focus on scientific issues, their criticisms of *AIT* and Gore were both scientific (Edelman, 2007; McIntyre, 2007; Nova, 2009b) and personal (McIntyre, 2008; Nova, 2009a; Watts, 2008). Crucially, these commentators had a (small) number of similarly critical climate scientists upon whose knowledge they could draw. Two of these scientists published critiques of *AIT* as part of a series in *GeoJournal* (Legates, 2007; Spencer, 2007).

This network of critical actors was akin to a scientific counterpublic attempting to challenge the hegemonic representation of climate change sought by *AIT*. They were a relatively small number of scientists with connections to other societal actors sharing a concern about the interactions between science, power and politics (Hess, 2010: 631). This is not to say that the counterpublic is any closer to the truth, or freer from external biases, than the dominant public, only that *AIT* and Gore provided important rallying points around which a counterpublic could coalesce (Jaspal et al., 2014). The substance of this counterpublic's criticisms is already well documented in the literature (Koteyko et al., 2013; Lahsen, 2013; Matthews, 2015).

One particular characteristic of these criticisms is focused on here; the way in which critics sought to disassociate the notion of climate-science expertise from the representation provided in *AIT*. Jaspal et al. describe this as the challenging of science 'by appealing to its norms' (2013: 383). They highlight a reader comment on climate-change articles on the *Daily Mail* website that 'distances Al Gore from "science", which is interesting in itself, as he is not actually a scientist' (Jaspal et al., 2013: 395). Of course, Gore does not overtly claim to be a scientist; however, as the linchpin of *AIT* Gore became a cornerstone for the social representation of climate-science expertise. The reader comment claims that 'Gore stood to gain hundreds of millions of dollars' if legislation were passed lowering carbon emissions (Jaspal et al., 2013: 395).

It is indeed the case that two years before *AIT* Gore co-founded an investment management partnership focused on sustainability issues (Generation Investment Management, n.d.), and that one newspaper report claimed that his 'green-tech' investments boosted his net worth from $2 million to $100 million between 2002 and 2012 (Leonnig, 2012). Whether or not these figures are entirely accurate, they highlight the importance of the social context that is given to Dewey's 'bare ideas', and in particular the contested boundary between content and context (Brown, 2009: 159).

Brown (2009: 160) notes that the 'social conflicts associated with genetic engineering do not invalidate the theory of the double helix'. Similarly, the financial interests of Al Gore highlighted in the *Daily Mail* comment do not invalidate the fundamentals of atmospheric physics. However, the comment highlights the fuzzy boundary between content and context in the public sphere, and how a questionable context can bring the content into question and destabilise representations of expertise. Citizens' willingness to accept or challenge climate-science expertise is to some degree dependent on their core values (Kahan et al., 2011). One can't please all the people all the time. However, even assuming that Gore's intentions in making *AIT* were of the best, his financial interest in sustainability investments was not necessarily a firm foundation for his emerging public status as a climate-change expert.

While helping to raise the profile of climate change, *AIT* seems also to have contributed to polarisation and strengthened the voices of what some may call an 'inconvenient public' keen on publicising 'inconvenient knowledge' related to Gore's presentation of climate science and his own role as the public face of climate change. The use of the film to increase 'public understanding' of climate change was thus at one and the same time a success and a failure, a miracle and a monster.

Conclusion

In this chapter we have outlined the role of *AIT* in creating a strengthened social representation of climate change; making the impersonal personal and the invisible visible. By many measures *AIT* was hugely successful, winning numerous awards, earning Al Gore a share of the Nobel Peace Prize and providing a springboard for a global campaign of public education and activism. Drawing on the work of Brown, we have shown how *AIT*'s focus on creating new audiences for climate-science expertise echoes Dewey's original call for science to be returned to the people as 'a refinement of commonsense inquiry' and not to remain an entirely unfamiliar way of knowing (Brown, 2009: 160). The film also echoes Dewey in providing an aesthetic, emotional communication of expertise, going beyond the persistent deficit model in climate-change communications that assumes that the absence of concern about climate change is the result of a lack of knowledge (Nerlich et al., 2010; Pearce et al., 2015). In many ways *AIT* provides a model for bringing scientific expertise into the public sphere.

However, mistakes were made. In particular, errors on scientific content should have been avoided. As Hulme noted in his study of Gore's questionable comments on Mount Kilimanjaro's glaciers, returning scientific knowledge to the people 'may destabilise knowledge as much as it may legitimise it' and public trust in provisions for quality assurance in evidence are key (Hulme, 2010: 322). This goes for social representations of climate-change expertise as much as it does for scientific representations of nature appearing in the peer-reviewed literature. Whether these mistakes had a significant bearing on public attitudes towards *AIT* is beyond the scope of this chapter. However, what we have shown is how social representations of expertise inevitably bring context to content, and a boundary between the two that is contested. In the case of *AIT*, Gore's position as a Democrat politician formed part of the film, perhaps making Republican-supporting viewers less receptive to the film's message. Counterpublics may seek to bring in other contexts as a means of contesting social representations. In the example above we show how Gore's financial interests were used as a means of discrediting the scientific content. For scientists, this may seem anathema, but is the kind of issue that requires attention when returning scientific expertise from academia to the broader society.

In its mix of the scientific, personal and political, *AIT* is perhaps best thought of as an ambitious, if flawed, experiment in science communication and in making climate change meaningful. It did so, whether consciously or not, by politicising climate change and reintroducing the human into previously apolitical representations of climate change (Jasanoff, 2010). While agreeing with the need for politics, not science, to bear the load of dealing with climate change, we note that one effect of *AIT* was to turn climate science into 'Al Gore's science', closely tied to a narrow range of policy options that were anathema to US conservatives (Sarewitz, 2011). We also note that if future engagement on climate change is to improve on the experience of *AIT*, those taking part must be open to engaging with publics that might be regarded as inconvenient just as much as with invited and convenient ones. Such engagement can be rewarding or frustrating to various degrees (Hawkins et al., 2014), something we have both personally experienced with diverse publics on the Making Science Public blog that we have edited throughout the duration of the research programme. However, such engagement should continue if there is to be any hope of social representations of scientific expertise becoming a source of moderation rather than polarisation. We cannot, and should not, seek to vanquish the monsters lurking under the public face of science, but we might be able to do a better job of taming them.

References

An Inconvenient Truth > take action (2006). *Climatecrisis.net*, 6 July. Retrieved 7 January 2016 from: https://web.archive.org/web/20060706041140/www.climatecrisis.net/takeaction/.

Armstrong, J. K., Urry, A., Andrews, E., and Cronin, M. (2016). An oral history of *An Inconvenient Truth*. *Grist*. Retrieved 28 July 2016 from: http://grist.org/feature/an-inconvenient-truth-oral-history/.

Bates, E., and Goodell, J. (2007). Al Gore: 'The revolution is beginning'. *Rolling Stone*, 28 June. Retrieved 1 February 2016 from: www.rollingstone.com/politics/news/al-gore-the-revolution-is-beginning-20070628.

Beattie, G., Sale, L., and Mcguire, L. (2011). *An Inconvenient Truth*? Can a film really affect psychological mood and our explicit attitudes towards climate change? *Semiotica*, 2011(187), 105–125.

Booker, C. (2015). The real 'deniers' in the climate change debate are the warmists. *Telegraph*, 1 August. Retrieved 5 January 2016 from: www.telegraph.co.uk/comment/11778376/The-real-deniers-in-the-climate-change-debate-are-the-warmists.html.

Brown, M. B. (2009). *Science in Democracy: Expertise, Institutions, and Representation*. Cambridge, MA: MIT Press.

Climate Reality (n.d.). [How you can help solve climate change.] Retrieved 7 January 2016 from: www.climaterealityproject.org/training.

Cohen, B., and Shenk, J. (2017). *An Inconvenient Sequel*. Los Angeles: Paramount Pictures.

Daily Mail Comment (2015). Daily Mail comment: Climate change and an inconvenient truth. *Mail Online*, 24 July. Retrieved 5 January 2016 from: www.dailymail.co.uk/debate/article-3170198/DAILY-MAIL-COMMENT-Climate-change-inconvenient-truth.html.

Delingpole, J. (2010). South Park: The most dangerous show on television? *Telegraph*, 3 May. Retrieved 31 March 2017 from: www.telegraph.co.uk/culture/tvandradio/7671750/South-Park-The-most-dangerous-show-on-television.html.

Dewey, J. (1938). *Logic: The Theory of Inquiry*. New York: Holt, Rinehart and Winston.

Dewey, J. (1989). *Freedom and Culture*. Buffalo, NY: Prometheus Books.

Edelman, B. (2007). Detailed comments on *An Inconvenient Truth*. *Watts Up with That* [blog], 4 October. Retrieved February 25 2016 from: http://wattsupwiththat.com/2007/10/04/detailed-comments-on-an-inconvenient-truth/.

Elsasser, S. W., and Dunlap, R. E. (2013). Leading voices in the denier choir: Conservative columnists' dismissal of global warming and denigration of climate science. *American Behavioral Scientist*, 57(6), 754–776.

Fletcher, A. L. (2009). Clearing the air: The contribution of frame analysis to understanding climate policy in the United States. *Environmental Politics*, 18(5), 800–816.

Generation Investment Management (n.d.). *AlGore.com*. Retrieved 27 July 2016 from: www.algore.com/project/generation-investment-management.

Gore, A. (2006). *An Inconvenient Truth: The Planetary Emergency of Global Warming and What We Can Do About It*. London: Bloomsbury.

Grundmann, R., and Scott, M. (2014). Disputed climate science in the media: Do countries matter? *Public Understanding of Science*, 23(2), 220–235.

Guggenheim, D. (2006). *An Inconvenient Truth*. Los Angeles: Paramount Classics.

Haag, A. (2007). Climate change 2007: Al's army. *Nature*, 446(7137), 723–724.

Hawkins, E., Edwards, T., and McNeall, D. (2014). Pause for thought. *Nature Climate Change*, 4(3), 154–156.

Hess, D. J. (2010). To tell the truth: On scientific counterpublics. *Public Understanding of Science*, 20(5), 627–641.

Hollin, G. J. S., and Pearce, W. (2015). Tension between scientific certainty and meaning complicates communication of IPCC reports. *Nature Climate Change*, 5(8), 753–756.

Hulme, M. (2009). *Why We Disagree About Climate Change: Understanding Controversy, Inaction and Opportunity*. Cambridge: Cambridge University Press.

Hulme, M. (2010). Claiming and adjudicating on Mt Kilimanjaro's shrinking glaciers: Guy Callendar, Al Gore and extended peer communities. *Science as Culture*, 19(3), 303–326.

IMDb (2015). *An Inconvenient Truth*: Awards. *IMDb*. Retrieved 16 December 2015 from: www.imdb.com/title/tt0497116/awards.

IPCC (2007). *Climate Change 2007: The Physical Science Basis*. Contribution of Working Group I to the Fourth Assessment Report of the Intergovernmental Panel on Climate Change. S. Solomon, D. Qin, M. Manning, Z. Chen, M. Marquis, M. Averyt, M Tignor, et al. (eds). Cambridge: Cambridge University Press.

Jacobsen, G. D. (2011). The Al Gore effect: *An Inconvenient Truth* and voluntary carbon offsets. *Journal of Environmental Economics and Management*, 61(1), 67–78.

Jasanoff, S. (2010). A new climate for society. *Theory, Culture and Society*, 27(2–3), 233–253.

Jaspal, R., and Nerlich, B. (2014). When climate science became climate politics: British media representations of climate change in 1988. *Public Understanding of Science*, 23(2), 122–141.

Jaspal, R., Nerlich, B., and Cinnirella, M. (2014). Human responses to climate change: Social representation, identity and socio-psychological action. *Environmental Communication*, 8(1), 110–130.

Jaspal, R., Nerlich, B., and Koteyko, N. (2013). Contesting science by appealing to its norms: Readers discuss climate science in the *Daily Mail*. *Science Communication*, 35(3), 383–410.

Kahan, D. M., Jenkins-Smith, H., and Braman, D. (2011). Cultural cognition of scientific consensus. *Journal of Risk Research*, 14(2), 147–174.

Koteyko, N., Jaspal, R., and Nerlich, B. (2013). Climate change and 'climategate' in online reader comments: A mixed methods study. *Geographical Journal*, 179(1), 74–86.

Lahsen, M. (2008). Experiences of modernity in the greenhouse: A cultural analysis of a physicist 'trio' supporting the backlash against global warming. *Global Environmental Change*, 18(1), 204–219.

Lahsen, M. (2013). Anatomy of dissent: A cultural analysis of climate skepticism. *American Behavioral Scientist*, 57(6), 732–753, doi: 10.1177/0002764212469799.

Legates, D. R. (2007). *An Inconvenient Truth*: A focus on its portrayal of the hydrologic cycle. *GeoJournal*, 70(1), 15–19.

Leonnig, C. D. (2012). Al Gore has thrived as green-tech investor. *Washington Post*, 10 October. Retrieved 27 July 2016 from: www.washingtonpost.com/politics/decision2012/al-gore-has-thrived-as-green-tech-investor/2012/10/10/1dfaa5b0-0b11-11e2-bd1a-b868e65d57eb_story.html.

McIntyre, S. (2006). Day four: Al Gore. *Climate Audit* [blog], 14 December. Retrieved 16 December 2015 from: http://climateaudit.org/2006/12/14/day-four-al-gore/.

McIntyre, S. (2007). Dasuopu versions. *Climate Audit* [blog], 12 January. Retrieved 25 February 2016 from: http://climateaudit.org/2007/01/12/dasuopu-versions/.

McIntyre, S. (2008). How Al Gore saved Christmas. *Climate Audit* [blog], 25 December. Retrieved 25 February 2016 from: http://climateaudit.org/2008/12/25/al-gore-saves-christmas/.

Matthews, P. (2015). Why are people skeptical about climate change? Some insights from blog comments. *Environmental Communication*, 9(2), 153–168.

Moscovici, S. (1963). Attitudes and opinions. *Annual Review of Psychology*, 14(1), 231–260.

Moscovici, S. (1973). Foreword. In C. Herzlich (ed.), *Health and Illness: A Social Psychological Analysis* (pp. ix–xiv). London: Academic Press.

Moscovici, S. (1981). On social representations. In J. P. Forgas (ed.), *Social Cognition: Perspectives on Everyday Understanding* (pp. 181–209). London: Academic Press.

Moscovici, S. (1988). Notes towards a description of social representations. *European Journal of Social Psychology*, 18(3), 211–250.

Moscovici, S. (2000). *Social Representations: Explorations in Social Psychology*. Cambridge: Polity Press.

Moscovici, S., and Hewstone, M. (1983). Social representations and social explanations: From the 'naive' to the 'amateur' scientist. In M. Hewstone (ed.), *Attribution Theory: Social and Functional Extensions* (pp. 98–125). Oxford: Blackwell.

Mr Americana (2015). High priest of global warming cult, Al Gore, training missionaries of climate change. *Overpasses for America*, 30 September. Retrieved 29 February 2016 from: http://overpassesforamerica.com/?p=16885.

Murray, R., and Heumann, J. (2007). Al Gore's *An Inconvenient Truth* and its skeptics: A case of environmental nostalgia. *Jump Cut*, 49. Retrieved 4 November 2015 from: www.ejumpcut.org/archive/jc49.2007/inconvenTruth/text.html.

Nerlich, B., and Jaspal, R. (2014). Images of extreme weather: Symbolising human responses to climate change. *Science as Culture*, 23(2), 253–276.

Nerlich, B., and Koteyko, N. (2009). Compounds, creativity and complexity in climate change communication: The case of 'carbon indulgences'. *Global Environmental Change*, 19(3), 345–353.

Nerlich, B., Koteyko, N., and Brown, B. (2010). Theory and language of climate change communication. *Climate Change*, 1(1), 97–110.

Nolan, J. M. (2010). *An Inconvenient Truth* increases knowledge, concern, and willingness to reduce greenhouse gases. *Environment and Behavior*, 42(5), 643–658.

Nova, J. (2009a). Climate money: Monopoly science. *JoNova* [blog], 31 July. Retrieved 25 February 2016 from: http://joannenova.com.au/2009/07/climate-money/.

Nova, J. (2009b). Carbon rises 800 years after temperatures. *JoNova* [blog], 14 December. Retrieved 25 February 2016 from:http://joannenova.com.au/2009/12/carbon-rises-800-years-after-temperatures/.

O'Hara, H. (2015). *An Inconvenient Truth* review. *Empire*. Updated 1 January 2000. Retrieved 29 January 2016 from: www.empireonline.com/movies/inconvenient-truth/review/.

Orlowski, J. (2012). *Chasing Ice*. New York: Submarine Pictures.

Parker, T. (2006). ManBearPig. *South Park*, season 10, episode 6, 26 April. Comedy Central. Retrieved 26 February 2016 from www.youtube.com/watch?v=4oNfEI06YkY.

Pearce, W., Brown, B., Nerlich, B., and Koteyko, N. (2015). Communicating climate change: Conduits, content, and consensus. *Climate Change*, 6(6), 613–626.

Rayner, S. (2006). What drives environmental policy? *Global Environmental Change*, 16(1), 4–6.

Reynolds, G. (2005). Presentation Zen: Beginnings. *Presentation Zen* [blog], 18 January. Retrieved 29 January 2016 from: www.presentationzen.com/presentationzen/2005/01/the_blog_begins.html.

Reynolds, G. (2007). Al Gore: From 'showing slides' to winning an Oscar. *Presentation Zen* [blog], 28 February. Retrieved 29 January 2016 from: www.presentationzen.com/presentationzen/2007/02/al_gores_inconv.html.

Sarewitz, D. (2011). Does climate change knowledge really matter? *Climate Change*, 2(4), 475–481.

Sharman, A. (2014). Mapping the climate sceptical blogosphere. *Global Environmental Change*, 26, 159–170.

Spencer, R. W. (2007). *An Inconvenient Truth*: Blurring the lines between science and science fiction. *GeoJournal*, 70(1), 11–14.

Steig, E. J. (2008). Another look at *An Inconvenient Truth*. *GeoJournal*, 70(1), 5–9.

Tufte, E. R. (2003). *The Cognitive Style of PowerPoint*. Cheshire, CT: Graphics Press.

Turnbull, A. (2011). The really inconvenient truth or 'it ain't necessarily so'. Briefing paper no. 1. London: Global Warming Policy Foundation. Retrieved 3 October 2012 from: http://50.116.89.150/globalwarming/wp-content/uploads/2012/08/Happer-The_Truth_About_Greenhouse_Gases.pdf. [Now available at: www.thegwpf.org/images/stories/gwpf-reports/lord-turnbull.pdf.]

Watts, A. (2006). Global warming on Mars? *Watts Up with That* [blog], 20 December. Retrieved 16 December 2015 from: http://wattsupwiththat.com/2006/12/20/global-warming-on-mars/.

Watts, A. (2008). Gore to press: Stay out! *Watts Up with That* [blog], 13 April. Retrieved 25 February 2016 from: http://wattsupwiththat.com/2008/04/13/gore-to-press-stay-out/.

Williams, T. D. (2017). Weather channel founder: Gore's 'Inconvenient Sequel' another 'scientific monstrosity'. *Breitbart*, 20 January. Retrieved 31 March 2017 from: www.breitbart.com/big-government/2017/01/20/weather-channel-founder-gores-inconvenient-sequel-another-scientific-monstrosity/.

13

'Science Matters' and the public interest: the role of minority engagement

*Sujatha Raman, Pru Hobson-West,
Mimi E. Lam, Kate Millar*

Much has been written about how the public are imagined and constituted in recent science–society developments. In this chapter we explore the relatively neglected but related question of how the relationship between science and the public interest is constituted. The question is timely in the wake of Britain's exit from the European Union and the election of Donald Trump as US president. Both have raised significant concerns about the future of public support for science, and of policymaking supported by scientific facts (see Introduction). These have spurred public mobilisation and reflection by scientists concerned about the implications for their profession (*Economist*, 2016), as well as for the public interest as a whole (*Guardian*, 2017). But when members of the public mobilise around scientific research or policy decisions involving science, how should we understand their relationship to the public interest? This is our focus in this chapter.

At the height of concerns over science–society relations, the then UK Prime Minister Tony Blair delivered a widely cited 'Science Matters' speech (Blair, 2002). This speech echoed wider criticism, which still continues in Britain and elsewhere, of public protest against topics such as genetically modified (GM) crop trials or animal experiments. In mobilising to articulate what are minority positions vis-à-vis 'public opinion' as a whole, such publics seem, at first, to represent a monstrous departure from the social order and, in turn, the public interest. Following the twin meanings of the figure of the monster (Haraway, 1992, and see Introduction), we will critically interrogate this assumption and illustrate how minority groups are capable of engaging with science in ways that allow alternative visions of the public interest to become temporarily visible and potentially compelling.

The 'Science Matters' speech provides an opening for our argument, which we develop in the context of two different cases of minority engagement with science. We first consider the case of activists campaigning against the use of animals in scientific research, who are effectively characterised in Blair's speech as opposed to the public interest (as well as to science). Blair also contrasted the UK situation with support for science elsewhere in the world, implying that how science and publics relate to each other in other countries is relevant for how the UK imagines these issues. We therefore also consider a case in Canada, where a minority indigenous group engaged with both science and a part of the state (the law) to overturn a policy decision in fisheries management. This second case shows that science and public engagements can sometimes constitute challenges to dominant *policies* held to be against the public interest, as well as constituting opposition to established ways of doing scientific research, as in the animal activists' case. Whether or not they succeed in overturning the status quo, the engagements of minority groups should be seen as central to periodic renegotiation of the social contract with science, innovation and wider public policies (Guston, 2000) through which the public interest is constituted.

Tony Blair issued a clarion call for ensuring that 'Government, scientists and the public are fully engaged together in establishing the central role of science in building the world we want' (Blair, 2002). With this statement Blair essentially invoked the principle that matters of importance to science and scientists are also matters of the wider public interest, which he was authorised to pursue as head of government. Yet, by signalling that the world 'we' want will not simply follow from the work of either science or government, his speech identified a key role for the public in this respect. Blair seemed to suggest that only when all parties worked together would it be possible to achieve the common goal of co-producing science and the public interest.

In effect, Blair's 'Science Matters' speech opened up a space for conceptualising engagement as a way of embedding the public interest in science. But how should we conceive of the public interest in scientific research in the first place? And what are the grounds on which the public might be expected to engage with scientists and the state in pursuit of such a shared interest? Blair's account reflected more broadly shared assumptions: first, the public already has an interest in science in advance of its engagement in it, and second, members of the public will engage mainly to support and secure this pre-given interest. The first represents the substantive aspect of science/public engagement (what is in the public interest), while the second captures the procedural or processual aspect (how the public interest is to be determined). The 'Science Matters' speech characterised the substance of the public interest in science mainly in terms of the ability of research to offer technological solutions to economic, health and environmental challenges. For Blair and like-minded others, a shared interest in these solutions called for a process of engagement by a majority public to limit the influence of minority critics. The potential of minority publics to stimulate periodic renegotiation of what is taken to be the public interest through their engagements with science is, however, missing from this picture. To explore this potential, we situate our analysis in the context of recent research in science–society relations.

Public engagement is a major theme in this research (e.g. Felt and Fochler, 2010; Marris, 2015; Mohr and Raman, 2012; Welsh and Wynne, 2013), where it has been explored in relation to the inclusion and exclusion of particular publics and perspectives, and imagined representations of the public and public opinion. How engagement relates to the pursuit of a shared public interest is often implicit in these discussions, but only rarely is it explored in its own right (but see Hess, 2011; Jasanoff, 2011; Wilsdon et al., 2005). In this chapter, we seek to fill this gap.

Callon (1994) was one of the first to draw attention to the question of how we might think about the relation between science and the public good, though this was in a discussion framed by economics rather than public engagement. Interest in a wider set of questions is emerging, however, with Helga Nowotny, following Yaron Ezrahi, recently calling for more sustained analysis of the relationship between science and the public interest (Nowotny, 2014). For reasons of space we bracket the different lineages of relevant overlapping terms; namely, public value, the public good or the common good. We use 'the public interest' to denote the dual meaning that, first, some matters involving scientific research are, in principle, *of interest* to members of the public and, second, that these matters affect what is in the best interests of the collective ('society'). The first meaning is captured in the Royal Society's (2006) report on science and the public interest, which emphasises a need to communicate research results to the public to help them understand how these impact on their lives and enable them to participate in debates of the day. This is interrelated with the second meaning, invoked by Nowotny (2014), of public or collective interest, which is commonly defined as distinct from private interest alone. This public interest may or may not extend to *all* aspects of research, but we cannot know what it covers in advance of concrete efforts *to engage with* publics or *by* publics already engaging with science matters. Likewise, the nature of this interest is varied, but we focus specifically on *political* interests in and about science. In addition to government, scientists and a general public, as spotlighted by Blair in his speech, we include other social actors and institutions in governance (Lam and Pitcher, 2012). These include organised civil-society groups and communities as well as other parts of the state; notably, the law. Following Mark Brown, the conception of politics we have in mind for this inquiry is of 'purposeful activities that aim for collectively binding decisions in a context of power and conflict' (Brown, 2015: 19). Importantly, this conception takes various modes of participation, including civil-society engagements in governance, as *part of* the pursuit of collective decisions that underlie institutions of representative democracy (see also Brown, 2009).

We begin by first developing a framework of five key principles distilled from science–society debates, where the public interest question has periodically emerged but lacked detailed scrutiny. We then illustrate the strengths and gaps of this framework through the two case studies of science and minority public engagement, one on animal research in the UK and the second on fisheries policy in Canada. In science and public engagement research these have been described in process-based language; so, 'unruly' publics are said to be disinvited or otherwise excluded from the collective (e.g. de Saille, 2015; Welsh and Wynne, 2013). While this might be true in particular contexts, such a framing in terms of inclusion/exclusion unwittingly detracts from full consideration of wider public-interest arguments that we raise here in our examination of minority publics engaging with science.

Conceptions of science, the public interest and conditions of engagement

How should we conceive of the public interest in science? In this section we first examine reasons why a common response to this question – namely, that science is intrinsically in the public interest – is inadequate, and why more socially embedded notions of accountability and the usability of science might offer a more nuanced response. We then examine limitations of the way accountability and usability are commonly framed, which in turn underline the need to consider how publics may engage with science. Finally, we ask: on what grounds might public engagement be expected to happen? Our response explains why engagement is a process not merely for opening up but also potentially for renegotiating the substantive question of what is in the public interest (see also chapter 1). For instance, renegotiating a social contract for ethical fisheries has been promoted as a way of managing and protecting fishery resources and other public goods (Lam and Pauly, 2010). In practice, both engagement and renegotiation may happen only rarely and cannot substitute for socially attuned forms of expertise, as Jasanoff (2003) has argued. But the *potential* to renegotiate remains crucial for times when established understandings of the public interest are called into question (Barnett, 2007).

The idea that science is intrinsically in the public interest has resonated in the different registers of economics, culture and politics. In economic terms, this is underpinned by the classic definition of a public good as non-rivalrous and non-excludable, and thus deliverable only by a public body, not the market. Scientific knowledge, it is argued, meets these requirements (Stiglitz, 1999). However, this abstract notion of a good that can be used by people other than the producers has been critiqued and qualified to clarify the actual conditions that make it more or less possible to fulfil these requirements in practice. For example, recognising the rise of partnerships with commerce, Callon (1994) argued that science can still contribute to a public good but only through the pursuit of diverse questions and approaches, alliances with different networks, and the ability to share knowledge and support new collectives. Judging by this criterion, Stengel et al. (2009) conclude that UK plant science lacks the qualities of a public good.

The cultural case for an intrinsic public interest in science is also widely resonant, though the definition of scientific culture is more elusive, resting typically on the capacity of individuals and society to appropriate science (Godin and Gingras, 2000). This could encompass both the appreciation of science as a cultural good and a more instrumentalist economic understanding, though both may be linked in practice. For example, the local authority in Nottingham aims to develop the city as a place of scientific culture through its STEMCity initiative, which links education, community engagement and local economic development, and which has become a springboard for an ambitious responsible research and innovation project (Nucleus Nottingham, 2016). But whether made in isolation or in conjunction, both cultural and economic arguments for science in the public interest ultimately rest on expectations of the wider engagement in and use of science.

The limitations of taking the public interest to reside intrinsically within science become especially evident in the context of political arguments for supporting research. Whether the state should support specific lines of research or research in general is obviously a political question with public implications and implicit value choices. Political arguments for the intrinsic value of science are often mixed up with economic, cultural and societal rationales, as in Blair's 'Science Matters' speech. But such arguments should be seen as 'the *commencement* rather than the *completion* of public policy' (Guston, 2000: 48; emphasis added) and, more generally, an invitation to public discussion on what kinds of research are worth supporting and why (Brown and Guston, 2009). Precisely because they involve matters of public interest, claims on behalf of, say, state-funded, private or do-it-yourself research in synthetic biology should all be open to wider debate in the public sphere.

Public support for scientific research in turn entails that a public voice be heard. This has been recognised and promoted in policy through codes such as the Universal Ethical Code developed in 2007 by the British Government Office for Science, and through initiatives in public dialogue in emerging research and technological fields. The ethical code refers to science's need for a social licence to operate, based on a continually renewed relationship of trust, resonating with the language of good governance, such as openness and transparency, briefly alluded to in 'Science Matters'. While these terms signify a concern with *accountability* in the sense of requiring research systems to give an account and take note of public responses, research councils have also emphasised a need to demonstrate the impact or *use* of research in practice. 'Science Matters' was primarily a plea to the public to engage with and support science rather than a plea for scientific accountability, but it emphasised the use-value of research in the form of technological benefits. Paralleling similar developments in the USA (Guston, 2000), notions of the intrinsic public interest in research have disappeared from the British social contract with science, which is now firmly centred on demonstrating accountability and usability through wider engagement with the public. But, in practice, the terms of this engagement have been too narrowly circumscribed. We thus set forward five key principles that we believe underpin science–society engagement in the public interest.

First, engagement is not the same as endorsement. Public engagement can indeed offer the possibility of enhanced public support for research, as Blair envisioned and as political theorists Brown and Guston (2009) argue. But this does not mean people will support a *specific* study or technological configuration as a result of public discussion, as 'Science Matters' implied. Informed scrutiny has the potential to open up substantive issues that may not have been anticipated in research systems. Engagement can lead to many alternative outcomes for the proposed research: enhanced public support, criticisms leading to modifications, or outright rejection (which might still be accompanied by the endorsement of *other* forms of research). All these outcomes represent collective efforts to construct what is in the public interest.

Second, engagement can generate learning by different parties. Public engagement can expand the scope of issues deemed relevant for discussion beyond those originally imagined. This could cover matters of governance on specific areas of research, but also ideological disagreements about the nature, structure and value of these investments. For example, de Saille (2015) found that the social-movement activists she interviewed were more sceptical about how research is regulated than about the research per se. Likewise, Marris (2015) argues that governance issues, such as the lack of transparency about commercial links, was a real concern for publics critical of synthetic biology research, not commercialisation per se or fundamental ideological objections (e.g. 'tampering with nature'). Others highlight the gap between research funded for commercial purposes versus its public value (Moriarty, 2008; Wilsdon et al., 2005).

Third, engagement can open up alternative pathways for research and innovation. In determining what should be supported in state-funded research, one must consider the specific area of research (such as its risks, benefits and value to specific parties) and opportunity costs – that is, what other possibilities are foregone (Brown and Guston, 2009). If diversity is a criterion of science as a public good (Callon, 1994), then the market fails as a mechanism for achieving this good, as it prioritises only what elites say we ought to want (Jones, 2013). Public engagement, properly understood and devised, might stimulate discussion not just of the merits of one research area, but the wider question of what kinds of research and innovation are needed to fulfil the public interest (Jones, 2013). For example, Hartley et al. (2016) argue that to properly assess the merits of an emerging technology, such as GM insects, due consideration must be given to alternative research pathways to address societal challenges.

Fourth, engagement may involve the use of science to open up alternative policy pathways. Public engagement with science can take different forms, ranging from the appreciation of scientific insights to employing (say) climate science to make a case for radical political, economic and social change, to opposing experiments using animals in research. This means engagement does not only refer to efforts initiated by research systems and policymakers – it can also emerge from below. Nor does it signify just technological goods as a marker of public interest. Publics may engage with what Jasanoff (2006: 24) calls 'public science', i.e. 'science that underwrites specific regulatory decisions, science offered as legal evidence, science that clarifies the causes and impacts of phenomena that are salient to society, and science that self-consciously advances broad social goals, such as environmental sustainability'. Public science may be used by governments or the law in support of specific decisions but it may also be used by publics appealing against or seeking to overturn such decisions to advance their interpretation of the public interest.

Fifth, as a summary principle, *engagement can help revivify what is understood to be in the public interest.* Public engagement is a process for opening up and potentially renegotiating what is in the public interest. In addition to efforts by policymakers to engage the public, engagement might also include mobilisation from below by publics seeking to engage on their own terms (de Saille, 2015), often providing 'the basis for publicity for an *alternative* view of the public benefit' (Hess, 2011: 630; emphasis added). Knowledge from some areas of public science may be used to scrutinise or call to account other research areas – for example, those on environmental sustainability. Such engagements may emerge from 'scientific counterpublics' (Hess, 2011), who form alliances across different organisations and sectors (including science, non-governmental organisations (NGOs), professional groups and sympathetic parts of the state) and claim to offer a better account of the public interest than that assumed by dominant actors. The example we consider below of the Haida Nation represents one such diverse alliance. But engagement with science might also come from smaller, less-networked groups who have yet to persuade and mobilise a larger alliance of actors, but nonetheless have substantive issues to raise, as we will explore in the animal experiments case.

Both our cases highlight the limitations of conceiving public engagement solely as a procedural exercise for capturing the majority position, as often painted by policy sponsors of dialogue activities (Mohr and Raman, 2012). Studies of public-engagement exercises sometimes unwittingly reproduce the process-oriented languages of inclusion of public perspectives or the exclusion of uninvited or unruly publics, making it harder to focus on engagement as a mechanism for negotiating and potentially renegotiating substantive issues around science and the public interest. We explore these issues in our cases of animal-research activism and indigenous communities, which, in opposing different aspects of the dominant order, can be viewed as 'monstrous' in Blair's 'Science Matters' terms, or as warnings of the limits of this dominant order (Haraway, 1992).

Challenging animal research ... and the 'monstrous' public

Animal research is a particularly illuminating case through which to consider the limitations of the 'Science Matters' representation of public engagement as a process for endorsing current research systems. Animal research is a high-stakes issue, particularly in the UK (Hobson-West, 2010). Some argue that using animals is not just *a* method of science, it is *the* method of modern scientific inquiry (Rupke, 1987), creating an animal–industrial complex (Twine, 2013). Social scientists have framed animal research as dependent on a tacit social contract between scientists, citizens and the state (Davies et al., 2016). We show in this section that public engagement with animal research has the potential to open up alternative understandings of what kinds of research are in the public interest. Such alternatives have not yet been successfully established. However, the capacity for research systems and embedded notions of the public interest to change in the future cannot be ruled out.

In the UK opinion polls are commissioned regularly by the Government and receive significant media coverage (e.g. Department for Business Innovation and Skills, 2014). Results of these polls have had notable impacts on policy – for example, with funders supporting initiatives to open up animal research (McLeod and Hobson-West, 2016), partly on the assumption that this is what polls show the public want. Empirical research has also shown how different actors in the debate – including researchers using animals and animal-rights charities – claim to be aware of, and actually responsive to, public-opinion polling (Hobson-West, 2010). For those conducting animal research, claims that their actions are in line with public opinion represents a kind of legitimisation strategy (Hobson-West, 2012), so that the polls themselves become a route to a 'social licence to research' (Raman and Mohr, 2014). But beyond national opinion polling, or critiques thereof, how should we conceptualise different forms of publics in the animal-research debate?

One way is to focus attention on how minority groups are sidelined, silenced or undervalued in the sphere of animal-research policymaking. One key minority perspective is that animal research should not continue, either for ethical (cruelty to animals) or for scientific reasons (the unreliability of knowledge and technologies generated through animal research). This abolitionist view is sidelined in several ways, including via the use of opinion-poll results. For example, in the press release accompanying the 2014 Mori Poll (Department for Business Innovation and Skills, 2014, no pagination), the Government stresses that 'a majority of the British public accept the use of animals in scientific (medical) research "where there is no alternative"'. It then mentions the 'myths' that still exist, thereby implying that those who are not in the majority are misled. The minority view is also more implicitly sidelined in policy statements, for example, via the claim from the UK Home Office (2015, no pagination) that 'We respect the fact that people have strong ethical objections to the use of animals in scientific procedures. [But] we have legislated so experimentation on animals is only permitted when there is no alternative research technique and the expected benefits outweigh any possible adverse effects.'

In these examples, opponents of animal research can be understood as unruly publics (de Saille, 2015) who challenge the status quo. In the UK the peculiar history of active (and sometimes violent) protest against animal researchers means that labels of extremism abound, including in law, where legislation to control animal-rights activities was bound up with a government response to terrorism. This fits well with Welsh and Wynne's (2013) category of the 'threatening public', where the threat is both literal, in the sense of violence, and metaphorical, in that animal research and the life sciences are tied to economic growth (Home Office et al., 2014). If animal research is constructed as a key to medical progress, then an abolitionist agenda is enormously radical. To return to this book's metaphor, being seen as not on the side of health or progress is monstrous – that is, almost inconceivable – as an aberration of logic or civility.

However, one limitation with the focus on inclusion and exclusion of publics is that it can unwittingly reproduce a fragmented, individualised version of the public. An alternative analytical approach, following Hess (2011), is to look for dominant and subordinate *networks* and, crucially, to explore how those networks construct the concept of the *public interest*. Applying this lens to animal research, we can identify a dominant, currently stable network consisting of government departments such as the Home Office, the pharmaceutical and research industry, and, arguably, some powerful research charities, such as those campaigning for more research into diseases like cancer. The subordinate network comprises animal-rights groups, some religious groups opposed to using animals, and scientists and funders involved in using or searching for alternatives to animal research. In other words, opposition to the use of animals in scientific research is no longer seen simply as the vision of an aberrant public (as suggested by Blair) but as a position embedded within a set of alliances.

Given the link made between animal research and medical and health progress, one of the key discourses of the subordinate network is that animal research is *not* in the public interest. This is achieved in several ways, including by questioning the dominant narrative that animal research is equal to medical progress. Critics point to examples where results seen in animals have not transferred to success in human trials, the fact that many diseases remain without cures and the relative lack of research into unexplored areas of science, such as into alternatives to use of animals (Hadwen Trust, n.d.). Others question what it means socially, culturally and ethically to live in a society that tolerates deliberately killing or harming some species. In short, as predicted by Hess (2011), this scientific counter-culture is offering an alternative vision of the public interest. This is very different to an analysis focusing only on public consultation, where animal-rights groups might be seen as representing or giving voice to certain groups of individuals, or, as is implied by some of the names of campaign groups, such as SPEAK (http://speakcampaigns.org/), giving voice to animals themselves.

If counter-movements such as animal-research critics are indeed articulating alternative visions of the public interest, then, rather than being monstrous in the negative sense described by Blair, we could perhaps see them in more positive terms, as calling attention to limits of established ways of doing research. Following Haraway (1992), what these counterpublics potentially demonstrate is that alternative visions of medicine and science are possible, and that the established order may one day be overturned through the formation of new alliances that come to represent new scientific and social norms.

Challenging fisheries policy through a coalition of an indigenous community, public science and the law

We now turn to the case of an indigenous community, the Haida Nation, which is asserting and renegotiating the terms of its government-to-government relationship, as established in numerous agreements with the Canadian Federal Government, in the management of marine resources in its traditional territories. Disputes over fishing rights between British Columbia, First Nations, and the Department of Fisheries and Oceans Canada (DFO) reflect a history of legalised colonial dispossession and loss of access by aboriginals to fish (Harris, 2009). We examine the ongoing herring fishery dispute between the Haida Nation and the DFO to illustrate how public science can be used and combined with alternative forms of knowledge, such as traditional ecological knowledge (Berkes, 2012), in policy disputes. As in the animal-research case, this example highlights the limits of equating engagement with endorsing the object of engagement. But unlike that case, it also shows that subordinate networks (Hess, 2011) are capable of expanding their base – in this case, through a coalition of the indigenous community, stakeholders, indigenous and ecosystem-based science, and the law – to renegotiate what is understood to be in the public interest.

Haida Gwaii (formerly known as the Queen Charlotte Islands) is an archipelago on British Columbia's northwest coast with a population of approximately 5,000 residents, both Haida and non-Haida. The islands are the ancestral and contemporary home of the Haida Nation, which claims aboriginal rights and title to the archipelago. The Supreme Court of Canada has recognised that the Haida Nation has a strong *prima facie* case for the aboriginal title to Haida Gwaii, so the Federal Government has a duty to consult the Haida people and accommodate their interests (*Haida Nation v British Columbia*, [2004] SCC 73). Herring has significant cultural value for the Haida and other British Columbia Coastal First Nations, particularly as spawn on kelp, which is a traditional source of food and trade for indigenous peoples along British Columbia's coast. As a forage fish, herring plays an important provisioning role in the ecosystem, feeding predatory fish, birds and marine mammals, as well as supporting commercial roe herring, spawn-on-kelp, and food and bait fisheries in British Columbia. Herring stocks in the Haida Gwaii major stock area declined to chronically low levels in the 1990s and have yet to recover, resulting in closures of the commercial roe herring fishery since 2003 and spawn-on-kelp fishery since 2005 (Jones et al., 2017). However, in recent years, there have been a number of disputes over the proposed reopening of the commercial herring fisheries.

The inclusion and public consultation of First Nations' communities is prominent in fisheries management, but it is typically presented as a right of these groups to present their *own* special interests (von der Porten et al., 2016). This narrow characterisation of voice is part of an equally limited understanding of public engagement as a process for merely consulting different groups and acknowledging their distinct perspectives. Again, drawing on Hess (2011), and as implied in Welsh and Wynne (2013), we instead consider the possibility that minority voices are capable also of articulating a *wider* public interest. Minority communities often build specific claims for change, intervention or the protection of nature based on a collective vision of shared values and purpose. The Council of the Haida Nation (CHN) has articulated traditional Haida values (CHN, 2007) that it believes are important for planning marine use and managing fisheries (Jones et al., 2010). Haida values of respect, balance, interconnectedness, reciprocity, seeking wise counsel and responsibility have been compared to scientific principles of ecosystem-based management (Jones et al., 2010). These community values, if meaningfully taken into account in the engagement process, may widen what constitutes the public interest (Lam, 2015). This possibility was initially subverted by the DFO's use of public science to support its case for reopening the commercial herring fishery in Haida Gwaii. However, transient alliances between the CHN and the fishing industry in 2014 and the law in 2015 successfully challenged the DFO's construction of what was in the public interest.

In 2015, the DFO consulted the CHN and conducted preseason stock assessments that provided the option for closing down the commercial herring fishery around Haida Gwaii (Jones et al., 2017). Despite this, the then Minister of Fisheries reopened the commercial roe herring fishery, which led to a legal challenge from the CHN. Public engagement had occurred and procedural requirements had been fulfilled, yet the substantive arguments and claim presented in the consultation with the Haida that their values had been infringed was overridden in the final ministerial decision. The CHN filed an interlocutory injunction to prevent the reopening of the herring fishery based on four key points: (1) the herring stocks had not sufficiently recovered to support the commercial fishery opening, disagreeing with the DFO's scientific assessment; (2) given the infringement of Haida rights and title, the DFO had not adequately consulted and accommodated with the Haida Nation; (3) the DFO had failed to develop an integrated herring management framework with appropriate rebuilding strategies; and (4) reopening the fishery contravenes existing negotiated management agreements between the Crown and the Haida Nation.

The Federal Court ruled in favour of the Haida Nation (*Haida Nation v Canada (Fisheries and Oceans)*, [2015] FC 290). Judge Manson challenged the Minister's weighing of the scientific evidence and the lack of meaningful consultation and accommodation with the Haida Nation, given the significance of herring to the community's culture and traditions. While the Herring Industry Advisory Board supported opening the fishery, the judge noted:

> The [United Fishermen and Allied Workers Union] UFAWU, who are an integral part of the commercial fishery, supports the Haida Nation's position, for the very reasons why this injunction is being granted:
>
> i) the need for a better and independent science review of the herring stocks;
> ii) lack of inclusive decision-making;
> iii) their own assessment of the state of the roe herring stocks;
> iv) respect for local First Nations' insights;
> v) a willingness to build a collaborative understanding of the state of the herring in the shared ecosystem. (*Haida Nation v Canada (Fisheries and Oceans)*, [2015] FC 290 at [59])

Judge Manson cited the potential for irreparable harm to the herring and to the Haida Nation, as well as the balance of convenience, which weighs the potential prejudice to all parties, including the Haida Nation, the DFO, commercial fishers and the public interest. He concluded that granting the injunction was 'very much in the public interest'.

Thus, an alliance of indigenous-community, scientific, stakeholder and legal actors effectively challenged the Canadian Government's approach, both to engagement and to the use of science in informing policy decisions. Alternative sources of public science, different framings of knowledge, the significance of uncertainty and the role of values in informing a precautionary approach to resource management all became visible and, at least temporarily, powerful.

Conclusion

The fisheries case illuminates how a minority community successfully co-produced, with the law and scientific knowledge, an alternative vision of policy in the public interest. By contrast, in the animal-research case, activists opposed to the use of animals in scientific research have not yet been successful in institutionalising an alternative vision of the public interest, but this is not to say that such an alternative is precluded in the future. Both cases call attention to thinking more deeply about the grounds on which public engagement with science contributes to the public interest.

Science–society scholars suggest that the potential to diversify scientific practice and engage with diverse stakeholders is crucial for science to achieve the public interest (Stengel et al., 2009). They distinguish public value from commercial value, raise the importance of diversifying forms of innovation and bring in matters of the governance of cutting-edge science. These are all important but they focus mainly on processes for including publics, omitting the substantive matter of what is understood to be in the public interest at any one time. We have argued that insofar as public engagement is not simply a process for endorsing current research and policy practices, we need to pay more attention to its capacity to further the periodic scrutiny and renegotiation of what kinds of research and wider public policies receive support. In conclusion, we reflect on the potential of science–public engagements to transform what is taken to be in the public interest.

In his 'Science Matters' speech, Blair suggested that science is vital to Britain's continued future prosperity and that different parties need to collaborate to oppose the small band of obstructionists who were acting against the general public interest. Blair's speech invoked the legacy of Newton and Darwin and described science as 'just knowledge', thus attempting to side-step the relationship between science and commerce. Littered with references to nano-scale robots, biomedical science, hydrogen power and what he called e-science ('big data' in today's parlance), his speech overwhelmingly focused on technological outcomes from research producing new knowledge of how things work and the capacity to transform these operations, all ultimately linked to economic and financial benefits alone. He did not discuss, for example, the scientific knowledge that was making visible previously unforeseen hazards of industrial activity or drawing attention to the limits of technological fixes to environmental challenges. Nonetheless, Blair was appealing to a commonsensical view of scientific research for the greater good recurrently invoked in public discourse – most recently, by journalists urging an extension of the fourteen-day limit on embryo research to ensure benefits from medical science (e.g. Harris, 2016). In this equation of science and the public interest, the public are represented primarily as beneficiaries.

Yet, in principle, Blair's intervention opened up the possibility of renegotiating how the public interest in science is imagined, articulated and constructed through interaction among various different actors. Rather than taking the public interest as already given, the reference to engagement suggests that the interface between science and the public interest can on occasion be opened up and politicised in the sense of being '*made into* a part of politics' (Brown, 2015: 18), at least until a new settlement is achieved. The case of the Haida Nation's role in overturning a ministerial decision in Canadian fisheries policy by an alliance with public science and the law suggests that such renegotiations may be possible, but are likely to remain fragile unless they are supported by wider coalitions. So far, action on animal rights does not appear to have been able to similarly overturn dominant understandings of research in the public interest. However, our analysis suggests that we cannot foreclose future changes to received understandings, which are entirely possible through new and unexpected configurations of activism, public science, the law and publics. Until then, it is important to cultivate attention to apparently monstrous voices that seem to be discordant with the dominant order but may transform it in the future.

In conclusion, our two examples obviously do not negate the larger challenge of limits to public expertise and capacity to engage and scrutinise either science or policy, let alone articulate diverse perspectives. This capacity is necessarily limited in complex societies (Jasanoff, 2003), where facts are the aggregation of multiple, often proprietary, sources and complex institutional arrangements (Turner, 2015). In this context, state resources must be devoted to building independent and distributed systems of public expertise to engage and scrutinise, especially, large-scale research and innovation systems of the kind highlighted in Blair's speech. Until these systems are developed and their ability to elicit diverse perspectives is valued as much as research and innovation itself, efforts to connect science and the public interest are incomplete at best.

Acknowledgements

Thanks to Sarah Hartley and Russ Jones for helpful comments on previous drafts.

References

Barnett, C. (2007). Convening publics: The parasitical spaces of public action. In K. R. Cox, M. Low and J. Robinson (eds), *The SAGE Handbook of Political Geography* (pp. 403–417). London: Sage.

Berkes, F. (2012). *Sacred Ecology*. New York: Routledge.

Blair, T. (2002). Science matters. Speech to the Royal Society on 23 May 2002. Retrieved 1 December 2016 from: www.ukpol.co.uk/2015/09/12/tony-blair-2002-science-matters-speech/.

Brown, M. B. (2009). *Science in Democracy: Expertise, Institutions, and Representation*. Boston, MA: MIT Press.

Brown, M. B. (2015). Politicizing science: Conceptions of politics in science and technology studies. *Social Studies of Science*, 45(1), 3–30.

Brown, M. B., and Guston, D. H. (2009). Science, democracy, and the right to research. *Science and Engineering Ethics*, 15(3), 351–366.

Callon, M. (1994). Is science a public good? *Science, Technology and Human Values*, 19(4), 395–424.

CHN (2007). Towards a marine use plan for Haida Gwaii. N.p.: CHN. Retrieved 1 December 2016 from: www.haidanation.ca/Pages/Splash/Documents/Towards_a_MUP.pdf. [Now available at: www.haidanation.ca/wp-content/uploads/2017/03/Towards_a_MUP.pdf.]

Davies, G. F., Greenhough, B. J., Hobson-West, P., Kirk, R. G. W., Applebee, K., Bellingan, L. C., Berdoy, M., et al. (2016). Developing a collaborative agenda for humanities and social scientific research on laboratory animal science and welfare. *PLoS ONE*, 11(7), e0158791, doi: 10.1371/journal.pone.0158791.

de Saille, S. (2015). Dis-inviting the unruly public. *Science as Culture*, 24(1), 99–107.

Department for Business Innovation and Skills (2014). Public attitudes to animal testing. *Gov.uk*, press release, 4 September. Retrieved 1 December 2016 from: www.gov.uk/government/news/public-attitudes-to-animal-testing.

The Economist. (2016). The European experiment: Most scientists want to stay in the EU. *The Economist*, 28 May. Retrieved 7 April 2017 from: www.economist.com/news/britain/21699504-most-scientists-want-stay-eu-european-experiment.

Felt, U., and Fochler, M. (2010). Machineries for making publics: Inscribing and de-scribing publics in public engagement. *Minerva*, 48(3), 219–238.

Godin, B., and Gingras, Y. (2000). What is scientific and technological culture and how is it measured? A multidimensional model. *Public Understanding of Science*, 9(1), 43–58.

Guardian (2017). 'Science for the people': researchers challenge Trump outside US conference. *Guardian*, 19 February. Retrieved 4 April 2017 from: www.theguardian.com/us-news/2017/feb/19/epa-trump-boston-science-protest.

Guston, D. H. (2000). *Between Politics and Science: Assuring the Productivity and Integrity of Research*. Cambridge: Cambridge University Press.

Hadwen Trust (n.d.). *Dr Hadwen Trust*. Retrieved 1 December 2016 from: www.drhadwentrust.org/about-us/whats-the-problem. [No longer available, but see www.animalfreeresearchuk.org/mission-vision-values.]

Haida Nation v British Columbia (Minister of Forests), [2004] 3 SCR 511, [2004] SCC 73. *CanLII*. Retrieved 27 April 2016 from: https://scc-csc.lexum.com/scc-csc/scc-csc/en/item/2189/index.do.

Haida Nation v Canada (Fisheries and Oceans), [2015] FC 290. *CanLII*. Retrieved 27 April 2016 from: www.canlii.org/en/ca/fct/doc/2015/2015fc290/2015fc290.html.

Haraway, D. (1992). The promises of monsters: A regenerative politics for inappropriate/d others. In L. Grossberg, C. Nelson and P. A. Treichler (eds), *Cultural Studies* (pp. 295–337). New York: Routledge.

Harris, D. C. (2009). *Landing Native Fisheries: Indian Reserves and Fishing Rights in British Columbia, 1849–1925*. Vancouver: University of British Columbia Press.

Harris, J. (2016). It's time to extend the 14-day limit for embryo research. *Guardian*, 6 May. Retrieved 1 December 2016 from: www.theguardian.com/commentisfree/2016/may/06/extend-14-day-limit-embryo-research.

Hartley, S., Gillund, F., van Hove, L., and Wickson, F. (2016). Essential features of responsible governance of agricultural biotechnology. *PLoS Biol*, 14(5), e1002453, doi: 10.1371/journal.pbio.1002453.

Hess, D. J. (2011). To tell the truth: On scientific counterpublics. *Public Understanding of Science*, 20(5), 627–641.

Hobson-West, P. (2010). The role of public opinion in the animal research debate. *Journal of Medical Ethics*, 36, 46–49.

Hobson-West P. (2012). Ethical boundary-work in the animal research laboratory. *Sociology*, 46(4), 649–663.

Home Office, Department for Business Innovation and Skills, and Department of Health (2014). *Working to Reduce the Use of Animals in Scientific Research*. N.p.: Home Office, Department for Business Innovation and Skills, and Department of Health. Retrieved 4 December 2015 from: www.gov.uk/government/uploads/system/uploads/attachment_data/

file/277942/bis-14-589-working-to-reduce-the-use-of_animals-in-research.pdf.
Home Office (2015). 2010 to 2015 government policy: Animal research and testing. *Gov.uk*, Home Office, policy paper, 8 May: Retrieved 1 December 2016 from: www.gov.uk/government/publications/2010-to-2015-government-policy-animal-research-and-testing/2010-to-2015-government-policy-animal-research-and-testing.
Jasanoff, S. (2003). Technologies of humility: Citizen participation in governing science. *Minerva*, 41(3), 223–244.
Jasanoff, S. (2006). Transparency in public science: Purposes, reasons, limits. *Law and Contemporary Problems*, 69(3), 21–45.
Jasanoff, S. (2011). Constitutional moments in governing science and technology. *Science and Engineering Ethics*, 17(4), 621–638.
Jones, R. (2013). *The UK's Innovation Deficit and How to Repair It*. Sheffield Political Economy Research Institute paper no. 6. Retrieved 1 December 2016 from: http://speri.dept.shef.ac.uk/wp-content/uploads/2013/10/SPERI-Paper-No.6-The-UKs-Innovation-Deficit-and-How-to-Repair-it-PDF-1131KB.pdf.
Jones, R., Rigg, C., and Lee, L. (2010). Haida marine planning: First Nations as a partner in marine conservation. *Ecology and Society*, 15(1), 12.
Jones, R., Rigg, C., and Pinkerton, E. (2017). Strategies for assertion of conservation and local management rights: A Haida Gwaii herring story. *Marine Policy*, 80, 154–167, doi: 10.1016/j.marpol.2016.09.031.
Lam, M. E. (2015). Opinion: Herring fishery needs integrated management plan. *Vancouver Sun*, 9 November. Retrieved 23 November 2016 from: www.vancouversun.com/technology/Opinion+Herring+fishery+needs+integrated+management+plan/11505147/story.html.
Lam, M. E., and Pauly, D. (2010). Who is right to fish? Evolving a social contract for ethical fisheries. *Ecology and Society*, 15(3), 16.
Lam, M. E., and Pitcher, T. J. (2012). The ethical dimensions of fisheries. *Current Opinion in Environmental Sustainability*, 4(3), 364–373.
McLeod, C., and Hobson-West, P. (2016) Opening up animal research and science–society relations? A thematic analysis of transparency discourses in the UK. *Public Understanding of Science*, 25(7), 791–806.
Marris, C. (2015). The construction of imaginaries of the public as a threat to synthetic biology. *Science as Culture*, 24(1), 83–98.
Mohr, A., and Raman, S. (2012). Representing the public in public engagement: The case of the 2008 UK stem cell dialogue. *PLoS Biol*, 10(11), e1001418.
Moriarty, P. (2008). Reclaiming academia from post-academia. *Nature Nanotechnology*, 3(2), 60–62.
Nowotny, H. (2014). Engaging with the political imaginaries of science: Near misses and future targets. *Public Understanding of Science*, 23(1), 16–20.

Nucleus Nottingham (2016). Nucleus deliverable 4.6: Nottingham field trip report. *Horizon2020 NUCLEUS Project*. Retrieved 1 December 2016 from: www.nucleus-project.eu/wp-content/uploads/2016/08/4-06-NUCLEUS-Field-Trip-Report-Policymaking-Nottingham.pdf. [Now available at https://issuu.com/nucleusrri/docs/4-06_nucleus_field_trip_report_poli.]

Raman, S., and Mohr, A. (2014). A social licence for science: Capturing the public or co-constructing research? *Social Epistemology*, 28(3-4), 258-276.

Royal Society (2006). *Science and the Public Interest*. London: Royal Society. Retrieved 1 December 2016 from: https://royalsociety.org/~/media/Royal_Society_Content/policy/publications/2006/8315.pdf.

Rupke, N. A. (1987). Introduction. In N. A. Rupke (ed.), *Vivisection in Historical Perspective* (pp. 1-13). Beckenham: Croom Helm.

Stengel, K., Taylor, J., Waterton, C., and Wynne, B. (2009). Plant sciences and the public good. *Science, Technology and Human Values*, 34(3), 289-312.

Stiglitz, J. E. (1999). Knowledge as a global public good. In I. Kaul, I. Grunberg and M. A. Stern (eds), *Global Public Goods* (pp. 308-326). New York: Oxford University Press.

Turner, S. (2015). *The Politics of Expertise*. London: Routledge.

Twine, R. (2013). Addressing the animal-industrial complex. In R. Corbey and A. Lanjouw (eds), *The Politics of Species: Reshaping Our Relations with Other Animals* (pp. 77-95). Cambridge: Cambridge University Press.

von der Porten, S., Lepofsky, D., McGregor, M., and Silver, J. (2016). Recommendations for marine herring policy change in Canada: Aligning with indigenous legal and inherent rights. *Marine Policy*, 74, 68-76.

Welsh, I., and Wynne, B. (2013). Science, scientism and imaginaries of publics in the UK: Passive objects, incipient threats. *Science as Culture*, 22(4), 540-556.

Wilsdon, J., Wynne, B., and Stilgoe, J. (2005). *The Public Value of Science*. London: Demos. Retrieved 1 December 2016 from: www.demos.co.uk/files/publicvalueofscience.pdf?1240939425.

Part IV

Faith

14

Faith

Chris Toumey

I used to be younger. In 1987 I conducted an ethnography of the creationist movement as my dissertation research. Wonderful it was to be in the midst of the granddaddy of science and religion controversies in the years when creationism packaged itself as scientific creationism. That experience filled my head with ideas about relations between science and religion.

A note to our European readers, including the British: yes, I realise it is beyond strange that in a major Western nation a large proportion of the population continues to challenge evolution on the basis of religious belief. I cannot explain that here. I can only note that this is an enduring feature of life in the USA. Creationism is not going to go away anytime soon. Our conservatives here insist on celebrating US exceptionalism, and then the exceptionalism they give us is hostility to evolution. Oy vey.

Several years after I finished my work on creationism, when I wrote my second book, *Conjuring Science: Scientific Symbols and Cultural Meanings in American Life* (Toumey, 1996), I used a figure of speech that I called 'science in an Old Testament style'. The chosen people knew that their God has some human attributes. After all, isn't he an old white guy with a long beard who is frequently angry at the disobedience of the people he favours? But they know him less by his personality than by powerful mysterious signs like pillars of fire, burning bushes and dreadful plagues. My point was that many non-scientists respect and appreciate science, but for the most part they know science only in terms of superficial symbols that can be just as mysterious, and that can be manipulated to conjure a misleading image of scientific authority.

Now, to extend my simile: in the New Testament, God makes himself (or is it herself?) much less mysterious by arriving among us in human form. Walking with us, talking with us, eating with us and dying with us. If we are truly in an age when science is becoming more open and more public than before, then let us say that we are approaching science in a *New Testament* style and that we know that this is what we aspire to nurture.

One more phase: when we poor humans knew the God of the New Testament better than the one of the Old, there arose a new complication. A hierarchy of popes and bishops emerged to shape and dispense our knowledge of God. Another improvement became necessary. Martin Luther and others showed 500 years ago that each of us could know God personally and directly without needing clerics and theologians to manage our knowledge of God.

Is there an equivalent to the Protestant Reformation in non-scientists' knowledge of science? Do we need experts, and if so how do we need them, and how much do we need them?

After I finished my work on creationism I usually avoided those issues. One reason why I originally enjoyed my work on societal and cultural issues in nanotechnology was because I thought there were no issues of origins and ontology in this area. There is no religious denomination that I know of that argues that atoms and molecules are unreal. But in 2007 Jamie Wetmore at Arizona State University showed me that there were indeed some issues of religious reactions to nanotechnology, and that they are important. I have circled back to questions of the sacred and the profane in science and technology. Nanotechnology is unlike evolution, for which I am grateful, but I am back to where I should be when I examine new questions of science and religion.

I have a reason for sharing those thoughts with you. It is an honour for me to introduce the chapters in the theme of faith. As I do so, I see our theme not as a stand-alone part separate from the previous three themes, but as a question and a problem tightly intertwined with the others. Issues of the sacred and the profane can also be issues of openness, publics and experts. My figure of speech is a way to appreciate the other three themes as matters of the sacred and the profane; meanwhile, our explorations of science and religion benefit from the insights in the other three themes. For example, you see that I ask how similar is the process of knowing science to the three-millennium process of knowing the Judeo-Christian God.

And so here we appreciate the two chapters in this part. First, in chapter 15, we have a critique by Fern Elsdon-Baker of research on creationism and evolution, and specifically of the assumptions and methods that shape definitions and measurements of creationist sentiment. The author shows us that large-scale survey research on public attitudes about creationism and evolution fails to capture the nuance and diversity of those attitudes. Creationism, especially, is more interesting than what we see in those surveys. For example, creationists frame their views as scientific (calling their programme creation science or scientific creationism), as opposed to presenting themselves as anti-scientific. This then raises a complicated question: if creationists say that their programme is scientific, then what do they think science is? Survey research does not capture any of these interesting problems. In fact, the author's extended critique steers us to the conclusion that large-scale surveys are terribly problematic in examining any issue that embodies nuance and complexity, not limited to creationist thought. All this implies that a proper understanding of creationist and evolutionary thought, especially among multiple publics, is going to need the kind of thick description that comes more from face-to-face ethnographic work than from large-scale surveys.

A parallel problem is a package of assumptions about the phenomena to be measured. Elsdon-Baker tells the reader that researchers' thinking is secular to a fault; also, that they believe that they need to construct a contrast between science and religion. This is a tricky problem. True, scientific thinking is rightly grounded in secular values, and science should not be expected to execute any religious agenda. But it does not necessarily follow that science and religion are intrinsically incompatible. One hopes that this fallacy is well known, and that serious scholars recognise it, and that they frame their research accordingly. One hopes, but the author shows that this fallacy unfortunately remains pervasive in research on public attitudes about creationism and evolution.

The third theme that guides this chapter is the observation that most of the work on creation versus evolution controversies has been done in the USA. From this it might seem as if issues of creationism and evolution in other nations are imports from the USA, 'uncritically consumed' in local circumstances, as Elsdon-Baker puts it.

On the contrary, these controversies are just as much situated and adapted in local circumstances as the US versions are situated in US culture. The author points out that European values are far more secular than US ones and that this has clear consequences for the way that creationism flourishes, or fails to flourish, beyond the borders of the USA. Evolution elsewhere is hardly ever considered the nexus of all evil that American creationists believe it to be. This is not to say that creationist and evolutionary thought are either reconciled or synthesised in other nations, but neither are they amenable to a 'clash of civilisations' approach.

Here the author corroborates the 2004 collection *Cultures of Creationism,* edited by Coleman and Carlin. That work tells us that the Institute for Creation Research (ICR, located in southern California) produces and exports a particular vision of creationism. That vision is well received in many places by conservative Christians, e.g. certain missionaries and their converts in Kenya and South Korea, but then the host cultures adapt the ICR message to local circumstances. One might say that these variations of creationist belief show that ICR's message undergoes a process of nuance and diversity, exactly as Elsdon-Baker suggests (but which ICR probably did not intend).

Which be the monsters here? Creationists who threaten science? Fictitious creationists who are fabricated to conform to secular prejudices? Or misguided research assumptions and methods that contain structural fallacies which have the effect of suppressing the nuance and diversity that are part of the reality of creationism?

Next comes chapter 16 by David A. Kirby and Amy C. Chambers, on religious judgements about science in the movies of the twentieth-century USA. Much has been written about science and religion, and also about science and film, but this chapter nicely balances those three elements. This is important because one's religious values and beliefs often shape one's attitudes about science and technology.

We can ask whether Catholic and Protestant officials really understood the science – or rather the movie science – that they judged, and also whether they overestimated the power of film to shape viewers' minds and morals. Whatever the case, this chapter shows us that certain Christian leaders felt that they needed to control film content, and later to merely evaluate it, especially cinematic depictions of science. Topics of sexually transmitted diseases, birth control, eugenics, evolution and psychiatry aroused their disapproval, which would then lead to government bodies that would channel religious sentiment into government censorship. It is a relief to me that Kirby and Chambers do not try to measure movie science against realistic science. That might have been tempting, but it would have distracted from the strength of their chapter.

Which be the monsters here? Movie monsters? Monstrous offenses to morality that come from movies about science? Or are they the self-appointed gatekeepers of our morality who decided which stories of science in the movies would meet their approval?

The chapters

The two chapters in this part enhance the theme of tensions between experts and publics. Kirby and Chambers show us that self-appointed experts intended to control the ways that movie-going publics think about science and morality. I dare say that, in the long run, these would-be experts discredited themselves by insisting on small-minded interpretations of science and religion, and by imagining that they controlled audience behaviour more than they really did. I like today's status quo in which religious authorities offer denominational guidance and commentary, and in which their publics can accept or reject those views. Note that this is very different from a programme of aspiring to control what publics see and think about science and religion.

Elsdon-Baker's chapter adds more to the question of experts and publics. The experts who measure and describe creationism in terms of large-scale survey research are unlike the censors in the final chapter, but here we see that with the best of intentions experts can inadvertently distort the descriptions of creationism and creationists which publics encounter when survey research appears in mass media.

In much the same vein, we see that the themes of openness and responsibility resonate in these two chapters. If openness is good, then experts who are responsible to no-one else are contraindicated, as a pharmacist might say. It is not always easy to say to whom one is responsible, but here we appreciate that being responsible to no-one is bad for openness, and this at a time when there are good reasons to embrace openness in relations between science and religion. To put it another way, in tune with my earlier comments, it is inadvisable to underestimate the virtue and the influence of the Protestant Reformation. If we are moving in the direction of relations between faith and science that approximate the themes of the Protestant Reformation – including the theme of expecting experts to justify the need for their expertise – then it is good for experts to be responsible for explaining and justifying their values and methods.

The theme of transparency runs parallel to the theme of responsibility. Imagine how different it might have been if the experts presented in these two chapters had been required to explain in detail to their publics how they made their decisions.

I thank the editors for inviting me to introduce the chapters in this part, and also the authors for raising the issues they put before us. My role in this volume reminds me that there will never be anything simple about issues of science and religion. True, there are some simplistic opinions, but none of them do justice to these topics. There have been times when I wanted to walk away from those issues because they required more wisdom or more time for reading, research and writing than I possess. Good on you, Elsdon-Baker, Kirby, Chambers and the editors, for insisting that we must not turn away from difficult issues of science and religion.

References

Coleman, S., and Carlin, L. (2004). *Cultures of Creationism*. Aldershot: Ashgate.

Toumey, C. (1996). *Conjuring Science: Scientific Symbols and Cultural Meanings in American Life*. New Brunswick, NJ: Rutgers University Press.

15

Re-examining 'creationist' monsters in the uncharted waters of social studies of science and religion

Fern Elsdon-Baker

The subject of a clash between scientific and religious world views is often repeated as a very real 'fact' in scholarly, policy and public discourse – with creationists being painted as the ultimate unenlightened monsters that threaten scientific, and by extension societal, progress. There is, so we are told, a real and inevitable clash between world views – one that within extreme iterations can only be negotiated by an outright rejection of either science or religion.

We have become so accustomed to this framing of the relationship between evolutionary science and religion that it is now a commonly accepted norm within media or scholarly representation of evolutionary theory. Routinely, these accounts of anti-evolutionary stances tend to be reductionist and rarely look beyond polarised epistemic extremes. As Toumey notes:

> It is common for enemies of creationism to dismiss it as a simple exercise in Biblical inerrancy. Human evolution faces opposition supposedly because it contradicts Genesis 1:27, and evolutionary chronology is thought to attract enmity because it cannot be reconciled with a period of six literal 24-hour days. (Toumey, 1993: 296–297)

The frequently repeated account in public-space discourse is then that Darwin's publication of *On the Origin of Species* caused significant public and theological backlash, and there has been an ongoing battle between rational scientific champions and creationist dragons ever since. The New Atheists are portrayed as only the latest incarnation of those who are brave enough to take up the champion's banner.

However, there is currently a significant gap in research, either historically or contemporarily, into *publics'* perceptions of the relationship between science and religion. It is important to note here I am purposely using the term 'publics', *not* the public. This is in recognition of the fact that there are multiple ways in which we can think about the 'public' or indeed that members of such 'publics' may identify themselves within different contexts. There is also an increasing recognition that the boundaries between expert, scientist or public can be, and have historically been, fuzzy. Indeed, an expert in one subfield of scientific research is a 'public' to another distinct subfield of research, even in some cases within the same discipline. We are all – however ensconced we may be in an ivory tower, media organisation or system of governance – in differing contexts, at one time or another, a member of the public.

This broader conceptualisation of publics is often missing from research in this field, which tends to focus more narrowly on certain communities. What little systematic social scientific work that has been done has tended to focus on the US debates concerning 'creationism'. Often, the more sophisticated research that has been undertaken has focused on distinct faith communities or those working within elite scientific institutions. Therefore, beyond the polar extremes of these debates we have no real idea of how the supposed clash between world views plays out in the day-to-day lived experience of wider publics, or the role of wider identity politics, or indeed geopolitics, in relation to the role of religion and science in society. Moreover, we have no real understanding of whether this is an issue that has any salience or meaningful consequences for publics at all.

In the interests of reflexivity and as it is integral to the argument within this chapter, I want to stress that I am myself a lifelong atheist, from an atheist and scientific family, and have for a number of years worked within science communication – specifically communicating evolutionary biology. Indeed, in 2009 I directed the Darwin Now project for the British Council, a multi-million-pound project that ran educational, public-engagement and academic activities related to the Darwin anniversaries in up to fifty countries worldwide. It is this experience of working to engage publics with evolutionary science across a range of cultural contexts that led me to examine in more detail what the social and cultural drivers behind the idea of a necessary clash between science and religion might be.

There are a number of key assumptions at the core of this clash narrative, the primary one being that science and religion are in some form of intractable conflict, especially when it comes to evolutionary science. This implicit assumption pervades the way in which much of the debate or research surrounding creationism and evolution is conducted. These implicit assumptions, and in some cases outright prejudices, in the way that data are collected or research is framed, indicate that epistemic assumptions within scholarly research may contribute to an inflation of scale and threat in the public, policy and media gaze of the phenomena we might refer to as creationism, both in broader international context and in the USA (Elsdon-Baker, 2009, 2015; Hill, 2014a). This in turn means that by using reductionist categories and making inflated assumptions about their salience to individuals, we are in effect creating creationists in the way we collect data (Elsdon-Baker, 2015). As Jonathan Hill observes with regard to his 2013 National Study of Religion and Human Origins, a nationally representative survey of 3,034 US adults:[1]

> When we carefully define the various possible positions, and measure the certainty with which they are held, both anti-evolution creationists and atheistic evolutionists turn out to be very small proportions of the total population. (Hill, 2014b, no pagination)

Unfortunately, some of the larger scale polling work that has been undertaken concerning public perceptions of evolutionary science and religion is as a result of the implicit assumptions implied by binary measures potentially skewed. This is in part due to the nature of the methodologies employed in large-scale polling or data collection. But it is also in part due to a lack of reflexivity or critical engagement with the individual researchers' own, or traditional disciplinary, assumptions; namely, that evolutionary science and religion are inevitably in conflict. It is evident from the issues framing in some of this research that there is a set of interlinked value judgements or implicit biases that may be at play within facets of the research data

---

1 For further details of this survey see the publicly available report at: https://biologos.org/uploads/projects/nsrho-report.pdf.

collection in this area of study. These assumptions and biases can be characterised by the following statements:

- Anti-evolutionism is always creationism and creationism is always anti-evolutionism.
- Creationism is a product of religious world views or theological concerns and is inevitable within societies that are religious.
    → Therefore, anti-evolutionary stances are always driven by strict adherence to religious world views, not by other social or cultural factors.
- Creationism as a mainstream world view is a real-world phenomenon that is on the rise and must be tackled.
- Creationism is a threat to evolutionary science or science more broadly.
    → Therefore, evolutionary science is increasingly under threat from creationism, which is driven by religious world views, ergo (evolutionary) science is inevitably under threat from religious world views or societies where they are predominant.
    → Ultimately, by extension, therefore, adherence to religious world views will inevitably lead to a necessary rejection of some or all facets of scientific world views, perspectives or research.

Some or all of the above statements might seem to some to be clear common sense or common knowledge. However, whilst we might see some or all of these statements as given 'facts' and obvious parts of the contemporary narrative surrounding the communication of evolutionary science, it is important to note that these statements are not based to any great degree on empirical data. Nor are they based on any form of multilayered analysis that allows us to build a systematic understanding of what might be happening at a public-space-discourse, group or individual level. Furthermore, by ignoring these different levels of analysis we are also ultimately making an implicit assumption that for an individual to self-identify as a creationist is always to self-identify as an anti-evolutionist. This might seem like a very counter-intuitive statement. However, both in a categorical sense – that is, in the way we collect survey or polling data – and also in the way that people may themselves use the label 'creationist' as an identity marker, there are indeed times when a creationist is not a 'creationist' in the anti-evolutionist sense. A good example of this ambiguity over language and identity is the Church of England's website. In an essay entitled 'Good Religion Needs Good Science' by the Rev. Dr Malcolm Brown, he states that:

> There is no reason to doubt that Christ still draws people towards truth through the work of scientists as well as others, and many scientists are motivated in their work by a perception of the deep beauty of the created world. (Brown, n.d., no pagination)

The term 'created world' is here used in the context of accepting Darwinian evolution as the process 'by which humanity came to be'. It is philosophically entirely plausible to accept evolution as a mechanism by which a meta-causal deity 'created' humans and all other life forms. Indeed, it is not unheard of, in my experience, for those who accept all facets of evolutionary theory and a 'God of first cause' to refer to themselves as creationists.

However, this ambiguity or complexity is not often reflected in the more binary examples of research data collection. Nor is it considered in the ways in which researchers or science communicators, who are often working within more 'secular' social-science settings or intellectual traditions, approach the entire issue.

Such implicit biases or value judgements concern me when thinking about how we might collect data that provides us with a clearer account of what the social and cultural drivers are for rejecting evolutionary theory. I will expand on this point by focusing, firstly, on the implications of the geographical focus and gaps in research data and methodological concerns in this field of study, and, secondly, on the philosophical issues relating to the researchers' own cultural, disciplinary and institutional context within the field of study itself.

Geography and anti-evolutionism

Beyond pockets of research undertaken in the USA there is a very significant lacuna in research of any kind that explores public perceptions or attitudes towards evolutionary science, let alone scholarly research that examines how broader publics perceive the relationship between their personal belief (or non-belief) and their attitudes towards evolution (Elsdon-Baker, 2015). The problem with this US focus in terms of both data and researchers is twofold. Firstly, there is a tendency to assume this subject has been fully researched by US-based researchers, which leads in places to rather naive extrapolations or generalisations from US research into public perceptions onto other culturally distinct publics. This is perhaps most notably problematic in parts of Europe where religion plays a very different role in public-space discourse. Here, non-religious, atheistic or indeed anti-religious narratives are far more prevalent and more likely to be the dominant narratives in these debates, which will, in turn, impact on the social or cultural factors at play. It also assumes that creationism is a US export that is uncritically consumed and rebroadcast by creationists from different belief systems worldwide. This entirely ignores the possibility that there may be a distinct set of localised social and cultural drivers relating to the rejection of facets of evolutionary science, so that what we may be observing are multiple types of creationism and ways of being a creationist.

Additionally, in some of the scholarly discussion to date there is an almost alarmist agenda or tone. This is becoming increasingly evident in discussions of creationism outside the USA, where there is a very significant lack of comprehensive data. This, in part, is an understandable response of those within broad science, technology, engineering and mathematics studies fields to a perception of an increase in creationist discourses outside US contexts (see Blancke et al. (2013) for a good example and comprehensive overview of this concern in Europe). While I am not denying that creationist or intelligent-design narratives and groups do exist in Europe and internationally, we need to be very wary – especially when we have little empirical data – of inflating the levels of acceptance, salience or influence of such positions. As Blancke et al. (2013) show, there have been some quite vocal incursions of advocates of creationism (or its fellow traveller, intelligent design) into policy discussions within European contexts. As is often the way in public debate or online forums, it is often those who shout the loudest who gain the most attention, but it would be folly for the academy to assume that being vocal or having access to a privileged platform means that such individuals are acting in a way that is representative of broader publics' actual perspectives.

Such a significant lack of information about what publics really think with regard to evolution, faith and belief carries the risk that both of the counter-oppositional sides of the polarised narratives within these debates – 'creationists' and 'hard-line atheistic evolutionists' – can make claims that cannot be countered or refuted in any way. This kind of abstract debate might seem to be of limited importance outside US educational-policy arenas. But into this research vacuum steps speculation, implicit bias and, in some cases, outright prejudice.

Within a broader geopolitical context, the notion of the 'threat of creationism' is being tied to more pernicious clash-of-civilisation narratives. In the UK, for example, there are some recent cases whereby this this kind of implicit bias might be implemented within UK educational policy.[2] This is particularly evident in relation to concepts of British values and the purported, yet to date empirically unfounded, threat of Islamic versions of creationism (Elsdon-Baker, 2015; Hameed, 2014).

Data, methods and rejection of evolution

All too often, given the gap in data, scholars are reliant on the few quantitative surveys that have been undertaken (e.g. Blanke et al., 2013; Gallup, n.d.; Hameed, 2008; Miller et al., 2006; Newport, 2014; Pew Research Center, 2013). Given the nature of the commissioning processes involved and the way in which this kind of polling is conducted, this kind of research builds in from the outset a framing of research questions that may contribute to an inflation of the numbers of people we might classify as creationist, which further serves to exacerbate the public and scholarly discourse on the inevitability of the clash between evolution on the one hand and creationism on the other (see Elsdon-Baker (2015) for a more thorough analysis of issues framing in quantitative research in this field). A core component of such surveys or studies tends to be the development of measures that

---

2 For an interesting example of this in relation to Nicky Morgan, then Education Secretary, see the article by Matthew Holehouse in the *Telegraph*, 7 August 2014: 'Toddlers at Risk of Extremism, Warns Education Secretary' (www.telegraph.co.uk/news/politics/11020356/Toddlers-at-risk-of-extremism-warns-Education-Secretary.html), where the teaching of creationism in nurseries is linked to 'extremism' as part of a counter-terrorist response to the Trojan horse affair.

seek to ascertain levels of rejection of evolution or acceptance of creationism. These are very much based on models that have been in use in the USA for the past few decades, and are evident in the Gallup polls that have been conducted since 1982, which ask:

> Which of the following statements comes closest to your views on the origin and development of human beings:
> 1. 'Human beings have developed over millions of years from less advanced forms of life, but God guided this process'.
> 2. 'Human beings have developed over millions of years from less advanced forms of life, but God had no part in this process'.
> 3. 'God created human beings pretty much in their present form at one time within the last 10,000 years or so'. (Gallup, n.d.)

The respondents who chose option 3, the 'creationist' option, have stayed relatively consistently at 40%–47% between 1982 and 2014 (Newport, 2014). However, as Hill (2014a) has shown, a systematic and conscious affiliation with creationism is more limited. When asked, only 8% of survey respondents agreed with all aspects of the conventional young-earth-creationist world view, involving the creation of earth in seven twenty-four-hour days and a historical Adam and Eve. A large proportion (37%) of those classified as creationists in Gallup's terms also do not see correct belief about evolution as 'very' or 'extremely' important.

While this kind of polling is indicative of broader trends, it cannot on its own give us an understanding of how salient these matters are to individuals in their day-to-day lives, or how other complex factors, including cultural identity, might play a role in how people respond to these questions. However, this kind of large-scale public-attitudes measure is relatively commonplace within research to date into public perceptions of science. Given the history, agendas and development of science communication or public understanding of science as an academic field of research (Bauer et al., 2007), and also significantly as a practice, in recent decades it is unsurprising that there has been little engagement with the growing body of work in the sociology of religion that seeks to understand groups' lived experience of beliefs, or approaches that might seek to understand belief systems as a form of cultural identity or a form of belief and belonging.

Public understanding and anti-evolutionism

So, while there has been an emphasis in scholarly approaches on the perceived rise in creationism and creationist discourses, or in bluntly measuring numbers of creationists, contextual research has been lacking. Such research would need to explore the salience of creationism across populations or cultural contexts, the real lived experience of such a phenomenon, and general levels of understanding of what key terms in polling or surveying research (e.g. evolution, ancestor, creator or god) might mean to the respondents. Furthermore, negligible research has been done outside the USA to examine creationist discourses themselves, or the impact and level of influence these creationist groups or discourses might have on the general population. Additionally, alongside such studies there is a need for more analysis of what 'creationist' public-space discourses might be responding to, for example, in terms of secular humanist discourses or the perceived (and real) atheistic framing of evolution and broader cultural-identity issues (Hameed, 2014; Toumey, 1993). These approaches would give us a much clearer idea of what is being rejected when individuals reject evolution in large-scale surveys or quantitative research. In order to do this effectively, we need to move away from simplistic models that assume that the only reasons one might reject evolution are due to a literalist religious stance, or that the debate is purely about a clash between epistemic world views (Elsdon-Baker, 2016; Evans and Evans, 2008; Toumey, 1993).

New approaches and their limits

Therefore, in order to really understand public perceptions of the relationship between religious or spiritual belief and science, or indeed perceptions of evolution per se, we should try to remove any form of value judgment when evaluating the relationship between creationist and evolution. As Wylie and Nelson similarly observe in their review of the impact of gendered contextual and epistemic values on scientific research: 'Typically, these effects arise and persist because scientists are unaware of the values informing the research traditions in which they are educated and in which they work' (Wylie and Nelson, 2007: 78).

However, drawing on the positive impact of feminist approaches in a range of fields including anthropology, biological sciences, archaeology and the history of science, it is clear that we do not have to retreat into relativism if we accept and recognise the epistemic contexts within which we work. As Wylie and Nelson (2007) further highlight, the enterprise of science itself has a shifting and changing agenda. This is both an opportunity and a concern, in terms of the highly politicised social-scientific study of religion, the critical reappraisal of the secularisation thesis, and the increasingly geopolitical or transnational understanding of science and religion debates within society. Particularly in light of feminist, or indeed postcolonial, approaches to science studies, which have not only sought to avoid disappearing gender or race, but have also provided a new lens through which to increase our understanding of certain fields.

And yet the problems for the social-scientific study of religion and science are manifold. Firstly, there is the real potential for a strong conflation between the perceived need for science communication, science education or scientific-practice-based research to draw a distinct demarcation between 'science' and 'non-science', and the way that researchers in these or related fields may tackle the boundaries between public perceptions of 'science' and 'religion'. Such concerns might lead to an obfuscation of the more fluid and complex ways that we define, practice or engage with that nebulous entity we call science. Furthermore, it also fundamentally disregards the possibility that 'science' can act as a cultural identity as well as a methodological framework for understanding the world around us.

Secondly, not only is there a case to be made that there is a disappearing of the epistemic facets of religious belief from the study of religion, but also there is an added issue relating to the development of the social-scientific study of religion as a 'secular' discourse or approach (Evans and Evans, 2008). This is largely due to the values-based contexts which in places, both historically and within the academy today, have at least defined social- or natural-scientific research in contradistinction to 'religious' world views, or at worst have framed science as openly anti-religious. In the latter polarising binary narrative, to be a scientist (in this case a social scientist) you must be non-religious. This is then further compounded by subtle and nuanced differences in terms of the context, disciplinary history, practice and methodologies between different cultural contexts in how religion (and science and religion) is studied academically, which is in evidence even between purportedly similar anglophone contexts, e.g. researchers based in the UK and the USA.

By identifying the ways in which an open-minded and reflexive examination of scholarly research in the social and cultural study of science and religion, as both a subject for empirical enquiry and a methodological commitment, we can potentially open up new innovative lines of enquiry. To achieve this the submerged biases within existing research need to be examined. If we expand Wylie and Nelson's argument beyond the gender biases, which have become increasingly more evident in the academy, to the possibility that there are secular or Western epistemic biases too, we will have to confront the agendas of science as a programme, a cultural identity and a geopolitical world view in order to create space for new, more fruitful avenues of empirical research which enable us to better understand the role of religious (or spiritual) belief in society and public perceptions of its relationship with science.

Furthermore, if we were to take a step back from some of the value judgements I outlined at the beginning of this chapter, we might be able to revisit the entire discussion of what public perceptions of evolutionary science are, and what might actually be acting as the cultural or social drivers that lead individuals or groups to adopt anti-evolutionary stances or to voice concerns or doubts over evolutionary science per se. Thus, if we move from simply focusing on 'anti-evolutionism' as a 'religious' reaction to evolutionary theory and start to unpick what might be wider moral and ethical, historical, cultural, or societal concerns, we will start to develop a more nuanced picture of what exactly is driving anti-evolutionary, and by extension broader, anti-science stances within certain groups or communities. This is especially pertinent for understanding public responses across international contexts where differences in colonial experience and legacy, or the current geopolitical framing of world views or cultural, religious or ethnic identities will undoubtedly influence the ways in which people form attitudes or perceptions of evolutionary science or indeed 'Western science' itself.

Moral values and anti-evolutionism

We therefore need to more carefully examine what exactly the moral concerns might be that lead to a rejection of, or lack of trust in, evolutionary science. Both historically and contemporarily there have, to say the least, been some highly socially objectionable misappropriations of evolutionary science, the most obvious being the 'scientific racism' and eugenics movement, which saw enforced sterilisations in a number of countries worldwide. This may seem like a mere historical concern that relates only to the interwar period and the atrocities of Nazi Germany. However, eugenical programmes were adopted by a number of countries across North and Latin America, Europe, and Asia (Broberg and Rolls-Hansen, 2005; Stepan, 1991; Stone, 2001). It should also not be forgotten that the last of these laws in the USA was not repealed until as late as 1979, and in Alberta, Canada, they were not repealed until 1972.

This is within living memory for many publics, individuals or communities. These programmes were not always aimed at those who were seen as 'mentally unfit', but there were also within a number of contexts (including in the US) programmes of enforced sterilisation that targeted indigenous, migrant or ethnic-minority populations. This can still leave a lasting legacy today, either in the current discourses, or perceptions, concerning ethnicity and genetic screening (for a good example of this in California, see Stern (2005)). More worryingly, forced sterilisation is still a contemporary issue for some ethnic groups, leading the World Health Organization (WHO) in 2014 to release an interagency statement seeking to tackle: 'forced, coercive or otherwise involuntary sterilisation', and noting that:

> in some countries, people belonging to certain population groups, including people living with HIV, persons with disabilities, indigenous peoples and ethnic minorities, and transgender and intersex persons, continue to be sterilized without their full, free and informed consent. (WHO, 2014: 1)

The linking of Darwinism, eugenics and creationism does indeed form a part of creationist anti-evolutionary literature. This is perhaps unsurprising given the fact that it was Charles Darwin's cousin Francis Galton who first coined the term eugenics, even though he did not do so until a year after Darwin's death in 1882 (it is, however, worth noting that Darwin and Galton strongly disagreed on mechanisms of inheritance). There are even attempts on websites such as creation. com to link Richard Dawkins and New Atheist discourses with earlier eugenically inspired approaches.[3] However, while there is evidence of concerns over eugenics in creationism literature and discourse, insufficient historical research has been undertaken to examine the possible links between the development of anti-sterilisation discourses and creationist discourses in North America or elsewhere. As Dack (2011) notes in his analysis of the Alberta eugenics movement and Sexual Sterilization Act:

> [A] number of pamphlets and books were published in Canada during the mid-1930s, mostly by religious organizations in Ontario and Quebec, which were widely circulated in Alberta and spoke out against the province's sterilization law. (Dack, 2011: 102)

If we take the case of enforced sterilisation in 1930s Alberta, Canada, as an example, we can begin to see how the impact of these laws and subsequent programmes could possibly be linked to the development of creationist anti-evolution discourses. There is, of course, a need here for further and systematic historical study.

Even within well-studied contexts such as North America, then, not enough research has been done that allows historical insights into the impact of social factors such as sterilisation laws and 'scientific racism' on consolidating public anti-evolutionary stances. What research has been done has tended (for sound methodological reasons) to focus solely on explicitly creationist discourses and does not tend to explore those on the periphery of those debates or who do not explicitly endorse a strong creationist stance. However, concerns over sterilisation and eugenics are cited in creationist literature as part of a whole package of moral issues that are linked to evolutionary theory (Numbers, 1993; Toumey, 1993). Within broader international contexts, there has been a long legacy of evolutionary theory being perceived as being entwined with racial stereotyping and reductionist anthropological approaches. This is, again, not just a historically situated concern, but has been

---

3 See http://creation.com/dawkins-and-eugenics.

prevalent in more recent highly problematic articulations of alleged 'racial' differences.

A classic example of this is the publication of *The Bell Curve: Intelligence and Class Structure in American Life* by Richard Herrnstein and Charles Murray in 1994, which controversially discussed racial differences in intelligence due to both genetic and environmental factors, with people of African descent being seen as inherently less intelligent than those of Caucasian or Asian descent. This is a view that was heavily echoed in the public scandal caused by the Nobel-Prize-winning geneticist James Watson in 2007.[4]

These more recent examples can only add to longer-term concerns within some groups or communities that facets of 'evolutionary science' or 'genetics' are inherently racist or sexist. More recently we have also seen this in the linking by Richard Dawkins in his rhetoric over 'Islamic creationism' with broader Islamophobic narratives (Hameed, 2014). Taking into account the issue of race and ethnicity alongside the broader issues of perceived links between 'evolution' and 'secularisation' will no doubt play an important role in any research that seeks to unpick increasingly complex and geopoliticised debates on Muslim perceptions of evolution. As Hameed (2014) states:

> The rejection of evolution may be becoming another contested marker for Muslim minorities [in Europe] in schools. In fact, one can see why evolution may be the target: for many, evolution is one of the prime reasons for the secularization of Europe. Furthermore, it is often pitted against religion in popular press and many conflate evolutionary biology with racism associated with Eugenics and social Darwinism. (Hameed, 2014: 393)

There is no causal relationship between adopting an evolutionary-science paradigm for research and some of the atrocities conducted under the banner of eugenics, social Darwinism or public policies of enforced sterilisation that have affected potentially hundreds of thousands of people due to their race, ethnicity, gender or mental health status. However, in public perceptions of evolutionary science these societal issues clearly do (rightly) count, and will most likely

---

4 See www.independent.co.uk/news/science/fury-at-dna-pioneers-theory-africans-are-less-intelligent-than-westerners-394898.html.

Re-examining 'creationist' monsters 273

for some individuals, groups or communities play a role in the way in which people might perceive evolutionary science today. This broader linking of moral or societal concerns and a mishmash of identity issues relating to secularism with evolutionary science is stressed in the work of Evans and Evans (2008) and builds on the work of Toumey (1993), who, when discussing the linking of evolutionary theory to secular humanism within US creationist discourses, argued that:

> The key to understanding the intellectual structure of creationist thought is to see that it is part of a larger body of thought, that is, fundamentalist moral theory. This latter body of theory addresses a broad range of issues and worries about modern American culture and social change. As a result, creationist commentaries on both creation and evolution often refer, directly or indirectly, to the moral meanings that make those issues and worries so urgent to creationists and other fundamentalists. Much of the existential content of creationist thought is a broad cultural discontent, featuring fear of anarchy, revulsion for abortion, disdain for promiscuity, and endless other issues, to which evolution is then appended. (Toumey, 1993: 297)

As Evans and Evans suggest, there may be factors beyond epistemological concerns at play, stating that 'public debates between religion and science are no longer about truth, but rather about values' (Evans and Evans, 2008: 100).

The influence of these kinds of social and cultural narratives upon the views held by broader populations globally about evolution or the relationship between science and religion has been understudied, as has the wider link between being anti-evolution and anti-science, or the levels of trust in science among individuals or groups who reject Darwinian evolution.

Unfortunately, due to the primary, and in some cases sole, focus of social scientists on a kind of literalist or religious fundamentalist creationism as the only show in town when it comes to reasons for rejecting evolution, these wider, more nuanced and, on occasion, entirely credible social concerns over the perceived application of evolutionary science in society or public policy are often lost. Added to this is the compounding factor that the individuals, groups or communities most likely to be impacted on by such discourses are often the least likely to have a voice in public science or education policy. Moreover, they are seldom seen as an important or primary audience for science communication initiatives. Lumping together these important – yet often complex and nuanced – moral, social, political and geopolitical concerns with what are caricatured as extreme fundamentalist religious positions inhibits us from building a better understanding of what might be at play when people say they reject evolution. Little or no data have been systematically collected across publics to discern how these factors may or may not impact on the ways in which people negotiate, accept or reject evolutionary science.

Future research directions

What is needed, then, is a more internationally focused programme of research into these kinds of social and cultural narratives that moves us away from a less reflexive or reductionist approach towards a contextual whole-systems approach which incorporates both quantitative and qualitative data collection. Indeed, I am currently directing one such programme of research with sociologists, social scientists, historians, social psychologists and myself as the token philosopher. We expect that this type of research will go some way towards allowing a multilayered analysis of the different social and cultural drivers for the rejection of evolutionary science across a spectrum of publics at the level of individuals, groups, communities and societies. Far from arguing that quantitative research plays no role in this more systematic approach, I am arguing that we need to recognise that the kind of survey work discussed above should not be used solely to give us one-off snapshots. Instead, it is important to examine in more depth the issues framing within question design, together with the cross-correlation between survey items, and to resist using simple measures to draw comparative conclusions across cultural contexts. This kind of survey research, when used in conjunction with more in-depth historical, social psychological and qualitative data, can provide us with an opportunity to understand trends within the broader societal narratives. Given the contextual and translation issues inherent in international research, we should be careful to view this as part of a package of cultural- or country-specific case studies, as opposed to some kind of pseudo-comparative league table, as has been in evidence in some of the literature in this field.

As Bauer (2008: 124) notes in his review of survey research and the public understanding of science, 'The abuse of survey data is clearly possible and documented ... But the misuse of an instrument does not exhaust its potential.'

There is therefore a clear need to develop a more systematic field of humanities and social-science research (incorporating expertise from the currently predominately distinct study of both 'science' and 'religion') that seeks to explore these questions and to ask: 'Is public-space religion always public-space science's ultimate "other"?' This is not only important in terms of our own research practice, but could enable a more nuanced understanding of the relationship between science and religion that provides space for a productive dialogue and avenues for public engagement across a range of issues relating to the role of science in society both nationally and internationally.

To return to the theme of this book: if we go looking for monsters real or imagined, we will often find them, as is the case with creationists. However, sometimes we should be aware that these are monsters only because we choose to see them that way. Furthermore, in terms of truly critically engaging with our own disciplinary norms, hegemonies and implicit biases (or indeed prejudices), it may well be that the monster we fear the most is the one we will find if we look too closely in the mirror.

Acknowledgements

With thanks to Stephen Jones, Alex Smith and Chris Toumey for their input and feedback.

References

Bauer, M. (2008). Survey research and the public understanding of science. In M. Bucchi and B. Trench (eds), *Handbook of Communication of Science and Technology* (pp. 111–130). London: Routledge.

Bauer, M., Allum, N., and Miller, S. (2007). What can we learn from 25 years of PUS survey research? Liberating and expanding the agenda. *Public Understanding of Science*, 16(1), 79–95.

Blancke S., Hjermitslev, H. H., Braeckman, J., and Kjærgaard, P. (2013). Creationism in Europe: Facts, gaps, and prospects. *Journal of the American Academy of Religion*, 81(4), 996–1028.

Broberg, G., and Rolls-Hansen, N. (2005). *Eugenics and the Welfare State: Sterilization Policy in Norway, Sweden, Denmark, and Finland*, 2nd edition. East Lansing, MI: Michigan State University Press.

Brown, M. (n.d.). Good religion needs good science. *Church of England* [website]. Retrieved 8 December 2016 from: www.churchofengland.org/our-views/medical-ethics-health-social-care-policy/darwin/malcolmbrown.aspx.

Dack, W. M. (2011). The Alberta eugenics movement and the 1937 Amendment to the Sexual Sterilization Act. *Past Imperfect*, 17, 90–113.

Elsdon-Baker, F. (2009). *Selfish Genius: How Richard Dawkins Rewrote Darwin's Legacy*. London: Icon Books.

Elsdon-Baker, F. (2015). Creating creationists: The influence of 'issues framing' on our understanding of public perceptions of clash narratives between evolutionary science and belief. *Public Understanding of Science*, 24(4), 422–439.

Elsdon-Baker, F. (2016). The compatibility of science and religion? In A. Carroll and R. Norman (eds), *Beyond the Divide: Religion and Atheism in Dialogue* (pp. 82–92). London: Routledge.

Evans, J. H., and Evans, M. S. (2008). Religion and science: Beyond the epistemological conflict narrative. *Annual Review of Sociology*, 34, 87–105.

Gallup (n.d.). Evolution, creationism, intelligent design. *GALLUP*. Retrieved 8 December 2016 from: www.gallup.com/poll/21814/evolution-creationism-intelligent-design.aspx.

Hameed, S. (2008). Bracing for Islamic creationism. *Science*, 322(5908), 1637–1638.

Hameed, S. (2014). Making sense of Islamic creationism in Europe. *Public Understanding of Science*, 24(4), 388–399.

Herrnstein, R., and Murray, C. (1994). *The Bell Curve: Intelligence and Class Structure in American Life*. New York: Free Press.

Hill, J. (2014a). *National Study of Religion and Human Origins*. Grand Rapids, MI: BioLogos Foundation. Retrieved 12 August 2016 from: https://biologos.org/uploads/projects/nsrho-report.pdf.

Hill, J. (2014b). The recipe for creationism. *BioLogos*, 2 December. Retrieved 12 August 2016 from: http://biologos.org/blog/the-recipe-for-creationism.

Numbers, R. (1993). *Creationists: The Evolution of Scientific Creationism*. Berkeley, CA: University of California Press.

Miller, J. D., Scott, E. C., and Okamoto, S. (2006). Public acceptance of evolution. *Science*, 313(5788), 765–766.

Newport, F. (2014). In US, 42% believe creationist view of human origins. *GALLUP*, 2 June. Retrieved 8 December 2016 from: www.gallup.com/poll/170822/believe-creationist-view-human-origins.aspx.

Pew Research Center (2013). Public's views on human evolution. *Pew Research Center*, Religion and Public Life, 30 December. Retrieved 12 August 2016 from: www.pewforum.org/files/2013/12/Evolution-12-30.pdf.

Stepan, N. L. (1991). *'The Hour of Eugenics': Race, Gender, and Nation in Latin America*. Ithaca, NY, and London: Cornell University Press.

Stern, A. M. (2005). Sterilized in the name of public health. *American Journal of Public Health*, 95(7), 1128–1138.

Stone, D. (2001). Race in British eugenics. *European History Quarterly*, 31(3), 397–425.

Toumey, C. (1993). Evolution and secular humanism. *Journal of the American Academy of Religion*, 61(2), 275–301.

WHO (2014). *Eliminating Forced, Coercive and Otherwise Involuntary Sterilization: An Interagency Statement*. Geneva: WHO Press.

Wylie, A., and Nelson, L. H. (2007). Coming to terms with the values of science: Insights from feminist science studies scholarship. In H. Kincald, J. Dupre and A. Wylie (eds), *A Value Free Science: Ideals and Illusions* (pp. 58–86). Oxford: Oxford University Press.

16

Playing God: religious influences on the depictions of science in mainstream movies

David A. Kirby, Amy C. Chambers

Research on public attitudes towards science has revealed that individuals' personal values and belief system are crucial factors in determining how they respond to new developments in science, technology and medicine, such as nanotechnology (Brossard et al., 2009; Nisbet and Scheufele, 2009; Scheufele et al., 2009; Toumey, 2011). Few cultural institutions have more influence on personal values and belief systems than religion, and few cultural products have as much impact on public perceptions of science as the mass media.

In popular works and in many scholarly texts the interface between science and religion has traditionally been depicted as one of unbridgeable conflict (Evans and Evans, 2008). This divide has a long pedigree in British Gothic literature. It takes early form in Mary Shelley's 1818 novel *Frankenstein* (Shelley 1998), where a scientist plays God and creates a grotesque creature, rendering himself monstrous in the making of what the outside world deems a monster. In Bram Stoker's *Dracula* (2003 [1897]), the fearless vampire hunters must turn to ancient religious rites to defeat a monster that has descended upon an unsuspecting and technologically advanced London on the cusp of a new century. A distrust of scientists, who have turned away from morality and religion to dabble disastrously in questions of creation, runs through classic science fiction stories of biological horror and hybridity, like H. G. Wells's *The Island of Dr Moreau* (1996 [1896]) and Robert Louis Stevenson's *The Strange Case of Dr Jekyll and Mr Hyde* (2000 [1886]), respectively.

Despite the cultural resonance of this conflict narrative, though, scholars have raised doubts in recent years about its historical and contemporary basis (Ferngren, 2002). Nonetheless, the relationship between science and religion remains a topic of concern for the scientific community as well for various religious communities. One of the spaces where these concerns play out is in the stories we tell about science in the mass media, which occasionally identify metaphorical monsters in the double sense of the word – *to warn and to show* – as outlined by Nerlich et al. in the Introduction to this volume.

In our research we find that religious communities, primarily Western mainstream Christians, have often attempted to influence the way stories about science have appeared on cinema screens because they believed that movies were a powerful force in determining our perceptions of the world. These religious groups were concerned about the ways that movies portrayed science's role in society and science's place as a knowledge producer, and tried to control how the stories were told and how audiences interpreted them. By examining the negotiations between religious groups and the entertainment industry, we reveal how the culturally powerful medium of cinema has historically served as an arena where science, religion and morality come into conflict.

In this chapter we will explore the ways that filmmakers have converted the sciences into cinematic products and how religious groups have altered, responded to and appropriated these scripts by formal and informal censorship, negotiations with filmmakers during production and distribution, and reviews written as guidance for religious audiences. This topic is far too large to be adequately covered in a single chapter. We can only provide a historical overview that focuses on Christian responses in the USA. This focus is justified by the fact that the USA has historically been the world's predominant producer of entertainment media and religious responses to movies have primarily emanated from Christian communities. The Christian community is not a monolithic entity, however.

This chapter will cover the diverse responses to science in movies among Christian groups, including differing responses from Catholics and Protestants. Through this exploration we provide some insights into what religiously minded people considered to be morally offensive, indecent, threatening or 'monstrous' about science and scientific ways of thinking. Religious responses to movie narratives show us the kinds of stories moral reformers did, and did not, want told about science as a social, political and cultural force.

1900–1933: origins of film censorship and movies as social propaganda

Religious anxieties about the moral impact of movies on the public began with the proliferation of nickelodeons in the 1900s. A number of reform organisations with religious orientations, such as the Women's Christian Temperance Union and the Federation of Churches, complained about the perceived immoral content of early movies. These groups worried that 'obscene' messages in films were degrading the morals of lower class and immigrant populations (Grieveson, 2004). Pressure from religious groups led to the creation of local censorship boards in municipalities and states across the USA in the 1910s. But the presence of local censors did not mollify religious protestors, who continually pushed for a government-administrated national censor board. Fearing this, the film industry established an autonomous self-censorship organisation called the National Board of Censorship in 1909 (renamed the National Board of Review in 1915). Although mainstream film producers agreed to submit their scripts to this board for approval, it proved to be ineffective, leading religious reformers to call for the creation of a federally run censorship board. Hollywood's response was to bring in Postmaster General Will H. Hays, who was also a Presbyterian deacon, in 1922 to head a new self-censorship organisation called the Motion Picture Producers and Distributors of America, which became popularly known as the 'Hays Office' (Black, 1996).

Religious groups' (and thus censors') concerns about the persuasive power of the cinema were in line with the thinking of contemporary social-science researchers about the influence of media messages on attitudes and behaviour, especially after witnessing the effectiveness of strategic propaganda during World War I. Activist groups of all types considered movies an ideal tool for social propaganda. These activists included public health officials, medical researchers and progressive reformers who used movies in campaigns to disseminate scientific discoveries about public health and to promote faith in scientific solutions to what were referred to as 'social diseases' such as syphilis, as well as other science-related social issues like eugenics and birth control (Parry, 2013; Pernick, 1996; Schaefer, 1999).

The producers of medical propaganda films believed that they were contributing to the moral good by persuading people to change their behaviour, but these films proved to be highly controversial. Initially, the difficulty for censors was that these films were all related to sexual reproduction and sex was the one subject that every censor board agreed was inappropriate for mainstream movies. But censors also considered that stories featuring modern medical science were emotionally upsetting and aesthetically unpleasant. Ultimately, responses to films dealing with venereal disease (VD), reproductive technologies and eugenics shaped subsequent national censorship policies by broadening censors' views on what aspects of a film were censorable beyond just sexual content.

Damaged Goods (1914) was the first motion picture to tackle the issue of VD, a term used until the 1990s, when it was replaced by the phrase 'sexually transmitted diseases'. The box office success of this sexually provocative morality tale resulted in the production of a host of other 'sex hygiene' films in the late 1910s, such as *The Spreading Evil* (1918) and *The Scarlet Trail* (1918). Despite their significant sexual content there was very little official censorship of these films because they endorsed morality and abstinence as the weapons for fighting VD (Schaefer, 1999). After World War I, however, censors' policies on VD films shifted dramatically when two government-produced educational films, *Fit to Fight* (1918) and *The End of the Road* (1918), were released to wider audiences. They differed from earlier films by focusing on the use of prophylactics as a method for reducing the spread of VD, which led to widespread condemnation by Catholic groups (Parry, 2013). In addition, one of the very first studies into the effect of the cinema on audiences concluded that VD films could be harmful to mixed-gender audiences and should not be shown indiscriminately (Grieveson, 2004). These studies, combined with the Catholic protests, forced censorship groups to re-evaluate the entertainment value of VD films and their appropriateness for public consumption. Ultimately, VD films spurred censors' construction of a distinction between entertainment and educational films. This distinction played a crucial role in later censorship policies and it led to the physical segregation of the places where these two types of film could be shown (Pernick, 1996).

The inclusion of prophylactics in VD films was a major issue for religious groups because they were concerned with any cinematic narrative depicting birth-control technologies. Contraceptives were illegal in the USA before 1918, but a large number of activists were working to repeal these laws. Movies became a battleground upon which both sides of the birth-control controversy tried to sway public opinion, with advocate Margaret Sanger's *Birth Control* (1917) competing with anti-birth-control films such as *The House Without Children* (1919). Unlike VD films, which escaped early censorship, birth-control films were heavily censored and often banned outright by state censor boards. In some cases birth-control proponents tried to get around religiously based censorship by promoting birth control as a better alternative to abortion, but this tactic was unsuccessful (Parry, 2013).

Religious organisations were not the only cultural group supporting censors' efforts to restrain public access to films featuring controversial medical topics like birth control. Medical scientists also strongly opposed activists' use of film. Birth control was a subject best left to scientific experts because its filmic depiction might undermine confidence in the medical professions by empowering the public to challenge medical authority. In this case, scientists joined religious reformers in endorsing the distinction censors made between entertainment and educational films (Ostherr, 2013).

Eugenics was one of the most controversial medical topics during this time and the subject appeared in a large number of propagandistic fictional films produced between 1910 and the mid-1920s, including the pro-eugenics *Heredity* (1915) and the anti-eugenics *The Regeneration of Margaret* (1916). Eugenics was a particularly thorny subject because it often led to overt discussions of birth control, sterilisation and euthanasia. In addition, religious groups considered these stories immoral because they portrayed human reproduction as an outcome of scientific tinkering rather than as the spiritual expression of matrimonial love. But censors targeted eugenics films not just for their sexual morality but also because they were intellectually demanding, emotionally upsetting and aesthetically unpleasant (Pernick, 1996). Many religious film viewers and censors believed that the images of deformed bodies were too distressing for most viewers and could even cause birth defects in pregnant women.

The desire to eliminate 'unpleasant' medical subjects provoked censors to go beyond merely policing sexual morality to enforce visual standards for movies. In this way eugenics and other medical films played a central role in the emergence of what Martin Pernick (2007: 30) refers to as 'aesthetic censorship'. Many of the informal censorship policies that had arisen in direct response to medical films were formalised by the later adoption of the Motion Picture Production Code. This meant that films that dealt with VD, birth control and eugenics had virtually disappeared from commercial theatres by 1930.

Although the Hays Office took a strong position on medical films, it ultimately proved to be as ineffective as the National Board of Review. Hays believed that the only way studios would abide by his office's recommendations was if they agreed to adhere to a formal set of guidelines as to what was censorable. In 1930 studio heads agreed to abide by a code of standards called the Motion Picture Production Code that had been written by two prominent Catholics (Leff and Simmons, 2001). Martin Quigley was editor of the trade paper *Motion Picture Herald*; Father Daniel A. Lord was a Jesuit priest. (We will use the term Production Code to refer to the Motion Picture Production Code of 1930.) The Hays Office, however, could not force studios to accept their suggestions. This meant that, despite their agreement to abide by the Production Code, studios still frequently ignored its recommendations (Olasky, 1985). The director of the Hays Office at this time, Colonel Jason Joy, took a particularly lax approach to the Production Code, which he viewed as a flexible set of guidelines rather than a hard and fast set of rules. Because of Joy's lenient approach, the period between 1930 and 1934 is referred to as the pre-Code era.

There were no specific policies addressing science in the Production Code, although the document did include language explicitly addressing previous issues related to eugenics, VD and birth control. Other aspects of medical science became censorable because they fell under the heading of 'repellent subjects'. The Hays Office warned studios about the potentially 'gruesome' nature of film sequences involving surgical operations (Lederer, 1998).

Science did run afoul of local censor boards' religious sensibilities during this period. The rise of the horror film caused a number of censorship problems, including concerns about the monstrous nature of modern science. Censors were concerned about the stories frightening audiences and the gruesomeness of monsters. But one of the primary issues was the blasphemous nature of plots in several films involving scientists usurping God's role as creator, including *Frankenstein* (1931), which several state boards banned for this reason. Censor boards also removed specific dialogue in which the scientists claimed to be 'playing God' in films such as *Frankenstein*, *The Invisible Man* (1933) and *Island of Lost Souls* (1932). The fact that the Hays Office did not remove these lines at the script stage indicates how lenient Joy's interpretation of the Code was before 1934.

1934–1966: controlling stories about science in the age of censorship

From the perspective of religious protestors, the Hays Office's failure to rigorously enforce the Production Code meant that movies were just as morally problematic as they were before its adoption. In response, Will Hays created the Production Code Administration (PCA) as a way to curtail calls by religious groups for a government censorship organisation (Black, 1996). This pressure also led the Catholic Church to form its own censorship organisation, the Catholic Legion of Decency, in 1933 (Walsh, 1996). (The organisation changed its name to the National Legion of Decency in 1935, but we will refer to it only as the Legion of Decency.) Tough-minded Catholic Joseph Breen took over as director of the PCA in 1934. Breen had the power he needed to force studios to alter their scripts to conform to the Production Code's standards, or he would withhold the PCA's seal of approval (Leff and Simmons, 2001). The Production Code consisted of twelve major categories: crimes against the law, sex, vulgarity, obscenity, profanity, costume, dances, religion, locations, national feelings, film titles and repellent subjects. As such, the PCA and the Legion of Decency exerted significant influence over the types of stories studios could tell about science.

The intersection between science and sex continued to be a problem for censors. The censors' ban on VD films, for example, nearly prevented

Warner Brothers from producing *Dr Ehrlich's Magic Bullet* (1940) about the scientist who discovered the first cure for syphilis (Lederer and Parascandola, 1998). Negotiations with the PCA ultimately led to a film that celebrates the scientist without any references to his science (Kirby, 2014). The PCA also routinely censored scripts that used science in conjunction with criminal activity, such as the deployment of scientific progress as a justification for criminal activity in *The Amazing Dr Clitterhouse* (1938), or the use of scientific methods to commit the 'perfect' crime in *Before I Hang* (1940).[1] But broad notions of blasphemy and indecency allowed the PCA to censor science under almost any of the twelve categories. A scientist manipulating the soul in *Captive Wild Woman* (1943) violated the category of religion, while the PCA removed a lab experiment performed on a former church altar in the unproduced 1951 script 'Green Light' under the category of locations.

The PCA considered certain scientific fields to be particularly problematic under the Production Code. Stories involving evolution were a constant issue for the PCA, especially after the controversial Scopes Trial in July 1925. *Island of Lost Souls* may have made it through the Hays Office unscathed during Jason Joy's directorship in 1931, but the inflexible PCA removed every evolutionary element when Paramount rereleased the film in 1941. The PCA also forced filmmakers to alter scripts for films such as *Dr Renault's Secret* (1942) because they believed discussions of Darwin and evolution would offend religious individuals.

While the PCA altered films before production, the Legion of Decency classified films after their completion. The Legion's film classification system to guide Catholic viewers about which films were suitable and which were questionable used three levels: A – morally acceptable, B – morally objectionable in part and C – condemned.[2] The Legion's judgement could seriously impact on a film's box office potential, so filmmakers were anxious to avoid a B or C classification (Black, 1996). Few films received C classifications for their scientific content, and

---

1 All information in this chapter concerning the PCA's censorship activities comes from the individual film files in the PCA files at the Margaret Herrick Library, Academy of Motion Picture Arts and Sciences, Los Angeles, CA.

2 All information in this chapter on the Legion of Decency's censorship activities comes from the individual film files in the Legion of Decency files at the Catholic University of America, Washington, DC.

those that did were either VD films such as *Damaged Goods* (1937), or films concerning reproduction like *Men in White* (1934).

Many films were given B classifications during this period because of the theological implications of their scientific depictions. For example, the Legion censured films that used scientific explanations to support the notion of transmigration of souls, as in *The Man with Two Lives* (1942) and *I've Lived Before* (1956). They also disliked film narratives that portrayed scientific progress as a more powerful progressive force than religion, as was the case with *Madame Curie* (1943). Like the PCA, the Legion took issue with films featuring psychiatry and evolution, but their responses to these depictions evolved over time along with Catholic policies. Before 1950 any depictions of evolution automatically led to a B classification. But the Legion embraced films with overt evolutionary themes like *Inherit the Wind* (1960) after Pope Pius XII acknowledged the Church's acceptance of biological evolution in his encyclical of 1950, *Humani generis*.

Fear of not obtaining the PCA's seal of approval or of receiving a B or C classification from the Legion of Decency led studios to appease these groups by altering their scripts or editing their final films. But filmmakers also took a number of other actions in order to get their scripts through the PCA or to avoid a B or C classification. The PCA often instructed filmmakers to consult the Catholic Church's Hollywood representative, Father John Devlin. Father Devlin's suggestions changed the scientific content of several films, including *Red Planet Mars* (1952), whose story originally involved a scientist perpetrating a religious hoax. Even before receiving formal feedback, studios would often consult the Legion of Decency or other religious groups as a means of proactively placating censors and smoothing the approval process, as was the case for the biopic *Freud* (1962), where the Legion provided advice on how to make this scientific story palatable to religious audiences. In the case of *The Beginning or the End* (1947), the filmmakers consulted extensively with Cardinal Francis Spellman, which led to overt religious overtones in a film about the development of the atomic bomb (Gilbert, 1997: 52). This means that modifications to cinematic stories about science often came not through censorship itself, but through the actions of filmmakers who were anticipating censure.

Despite the power of the PCA and the Legion of Decency, many religious organisations did not support the idea of movie censorship, even in the 1940s when the PCA and the Legion of Decency were at their most powerful. The Protestant Motion Picture Council (PMPC), for example, felt that censorship was morally reprehensible. Instead, they provided faith-based reviews in the *Christian Herald* that guided viewers but allowed the public to make their own decision about a film (Linnell, 2006).[3] The PMPC's reviews were not exclusively about morally problematic science in cinema. Reviews also covered stories about science that they found inspirational or that they believed reflected their value system. Unlike the PCA and Legion of Decency, the PMPC celebrated films about psychiatry, including giving a Picture of the Month award to Alfred Hitchcock's *Spellbound* (1945). They also embraced films featuring brave scientists undertaking expeditions in the pursuit of scientific progress, such as *Scott of the Antarctic* (1948), which they also named Picture of the Month. Ultimately, the PMPC preferred stories in which the goals of science aligned with the goals of religion by improving the human condition, as in *Sister Kenny* (1946).

Unlike the PCA and the Legion of Decency, the PMPC trusted audiences to make the 'correct' interpretations about science in cinema. Proponents of censorship like the PCA, however, felt that it was a better strategy to modify movie plots in order to tell what they considered more appropriate narratives about science. These differences in approach reflected differing attitudes to morality between Catholics and Protestants in the 1940s and 1950s. The Catholic Church dictated its conceptions of morality to its followers, who were then expected to adhere to these judgements. Protestants offered guidance but wanted people to make their own choices about morality (Curran, 2008).

The threat of censorship during this period forced filmmakers to make decisions about which science to include or remove, based on reasons that had nothing to do with artistic merit, as they anticipated censure. The censors' sense of moral certainty did not require them to even understand the science upon which they were passing judgement. Ultimately, the PCA and the Legion of Decency began to lose their influence in the 1960s owing to broader cultural changes, including

---

3 All information in this chapter on the PMPC comes from the individual film reviews in the *Christian Herald*.

the rise of television, an increasingly permissive social stance towards sexual matters and a more socially progressive attitude in the Catholic Church (Leff and Simmons, 2001). The PCA became less worried about the theological implications of science and refocused their efforts on retaining some influence over the growing depiction of graphic sex and violence. But concerns among religious groups about scientific content in films remained after the end of official censorship. Without the power to censor movies, however, these groups had to find other ways to influence the way that audiences interpreted cinematic stories about science.

1967–1992: new Hollywood and new science require new approaches

The Classification and Rating Administration (CARA) replaced the PCA in 1968. CARA advised and negotiated with studios over proposed movie content in order for a film to get its desired rating, but it did not censor content. Hoping the new system would increase audiences owing to the production of more ambitious films with uninhibited themes, the industry received the introduction of ratings warmly (Wyatt, 2000). Filmmakers, freed from the prohibition or restriction of material that they had endured under the religiously constrictive Production Code, created an adventurous and vibrant cinema (Neale, 2005). Science played an important role in this period, as immediate post-censorship Hollywood movies positioned controversial science and scientific ideas at the core of their narratives. The shift from censorship to ratings influenced the ways religious groups responded to the film industry.

Film reviews became one of the primary Christian strategies for dealing with Hollywood's perceived onscreen depravity. The Catholic church dissolved the Legion of Decency in December of 1965 and established a new movie oversight organisation named the National Catholic Office for Motion Pictures (NCOMP). Like the PMPC in the 1940s, the NCOMP decided to provide guidance at the point of reception rather than attempting to censor material prior to release (Gillis, 1999; Romanowski, 2012). The NCOMP's bimonthly *Catholic Film Newsletter* provided reviews through the lens of Catholic values, including their assessment of scientific content. The NCOMP was particularly sensitive to the deification of science and scientists in the films of the 1970s. They believed that films like *The Andromeda Strain* (1971), *Zardoz* (1974) and *The Terminal Man* (1974) 'worshipped' science and technology and apparently attempted to demythologise God.[4] The NCOMP also found recurrent science versus religion narratives to be problematic. For example, Catholic reviews were unhappy that religion was framed as the antithesis to science in *Planet of the Apes* (1968). Even films that depicted religion as morally right, such as *The Exorcist* (1973) and *A Clockwork Orange* (1971), still placed religion in opposition to science and this caused the NCOMP concern.

Another approach that religious groups took to controlling movie content after the censorship era was the introduction of film awards. Religious groups used film awards as way of praising the film industry when it produced films that they felt aligned with their religious values, and believed that these awards would encourage the production of more films with appropriate moral principles. The NCOMP launched annual film awards in 1965. Some of the earliest awards were given to science-based movies. The organisation awarded the 1966 Film of the Year to the controversial *The War Game* (1965), which was about the impact of a nuclear war and atomic science policy in Britain. In 1969 the National Council of Churches and the NCOMP awarded Stanley Kubrick's *2001: A Space Odyssey* (1968) their Film of the Year as part of a joint award programme. It also won the NCOMP's Film of Best Educational Value that year. Religious groups readily interpreted Kubrick's science-based film as a religiously valuable film because of its dealings with supreme beings, whether metaphorical, alien or divine. They hoped that these awards would encourage studios to produce science-based films that allowed for discussions of the nature of the divine and promoted a role for morality in scientific progress.

Filmmakers' post-censorship freedom allowed them to tackle more serious science-based topics. Humanity's stewardship of the earth became a prevalent theme; a concern also shared by religious communities during this time. The growing environmental movement inspired eco-films like *Silent Running* (1972), *Omega Man* (1971),

---

4 All information in this chapter on NCOMP's activities comes from the individual film files in the NCOMP files at the Catholic University of America, Washington, DC.

and *Soylent Green* (1973). Protestant publications like the *Christian Century* suggested that the church should be more active in the environmental movement and that religious groups must rethink their traditional attitudes to reproduction (Cobb, 1970). This attitude was reflected in their reviews of eco-films that celebrated nature but warned against humanity destroying creation (Kavanaugh, 1971). In the eco-horror *Soylent Green*, where starving humans unknowingly eat processed human remains, the church survives as a refuge for the masses and attempts to treat those whom science has failed. This was a theme that the NCOMP's reviewers found 'consoling'. Religion and faith became frequent elements of science-based films throughout the latter part of the twentieth century, appearing not only in opposition to science, but also as its ally.

Religious communities may have lost their direct input into film productions (via script and final approval) but there was still open dialogue between filmmakers and religious communities. Although this was an era of cinematic experimentation, many filmmakers continued to court religious audiences. Audiences would be quick to associate Charlton Heston of the biblical films of the 1950s with his title role in late 1960s and early 1970s dystopian narratives. Reviews of *Planet of the Apes*, *Omega Man* and *Soylent Green* pointed out that it was Moses fighting apes, humanoids and evil corporations in these cinematic futures.[5] Heston's casting allowed filmmakers to court traditional and religious audiences. Studios also supported the production of viewing guides, including those published by the Lutheran Church. There was even a Lutheran Church Study Guide created for the religiously controversial movie *The Exorcist*, which other Protestant groups, such as the Methodist Church, asked to use after the film's release.[6]

Filmmakers in the 1960s and 1970s also continued to work directly with religious organisations when their films dealt with sensitive topics. For example, director William Friedkin was in constant correspondence with the Roman Catholic Church in the USA throughout production

---

5 These film reviews can be found in the clippings files at the Margaret Herrick Library, Academy of Motion Picture Arts and Sciences, Los Angeles, CA.
6 See the Lutheran Study Guides Folder in the William Friedkin Papers at the Margaret Herrick Library, Academy of Motion Picture Arts and Sciences, Los Angeles, CA.

of *The Exorcist*, discussing the technical correctness of the religious rituals and the church's attitudes towards scientific practice.[7] Friedkin's consultation with the Church meant that, despite erroneous reports of Catholic outrage in the popular press, the NCOMP's response to the film was mostly positive because they appreciated the film's 'salutary reflections on religious belief and the limits of science' (NCOMP, 1974).

Many science fiction films released between 1968 and 1977 were dystopian and serious, drawing upon imagined science and futures that would see the end of humanity. But the unexpected success of a 20th Century Fox release in 1977 signalled a significant shift in the depiction of science and the future. *Star Wars* ushered in a new genre, and the era of science fantasy, as George Lucas termed it, began. *Star Wars* rejected the scientific realism that had defined science-based movies since 1968 by positioning itself firmly within the fantasy genre. The movie was well received by religious groups as a 'breath of fresh air' that avoided the unsettling science that had defined the science fiction of the 1960s and 1970s (Siska, 1977: 668). Comments from some of the film's producers backed up this religious reading. *Star Wars* producer Gary Kurtz, for example, told the First Congregational Church in Los Angeles that the film was a parable and that the spiritual nature of the characters and the notion of the Force were intended as touchstones for a predominately Christian audience (quoted in Dart, 1978). Other science fantasy movies of the late 1970s and early 1980s, including *Close Encounters of the Third Kind* (1977), *E.T. the Extra-Terrestrial* (1982) and *Back to the Future* (1985), offered religious groups little to worry about with their blockbuster, family-friendly focus.

1993–2015: courting religious audiences with reconciliatory narratives

Throughout the 1980s, religious groups continued to focus most of their attention on Hollywood's predilection for violence and sex rather than scientific content. But with the release of *Jurassic Park* in 1993 the religious community took a renewed interest in cinematic science. The film's success resulted in a subsequent flood of science-themed

---

7 See the William Friedkin Papers.

films that has not diminished (Kirby, 2011). Film reviews continued to be an important way for religious groups to respond to the science in movies. The rise of the internet in the 1990s increased the number of outlets for these reviews. But filmmakers were also beginning to appreciate the growing economic power of the Christian community. This awareness not only encouraged Hollywood filmmakers to court religious audiences for their science-based movies, it also convinced the Christian film industry that their own science-based movies could find success in mainstream theatres.

The high profile of *Jurassic Park* and the prominence of its evolutionary themes led to an almost unprecedented response from conservative Christians, who sought to blunt or reframe the film's scientific messages. Conservative Christian film reviews consistently deplored the film's overt discussions of dinosaur and bird evolution, with one reviewer calling it an 'unceasing barrage of evolutionist propaganda' (Dickerson, 1993). Several conservative Christian groups even tried to counter the film's pro-evolutionary stance by producing booklets and pamphlets explaining creationism and the 'real' origins of *Jurassic Park*'s dinosaurs (see figure 16.1). But it was the emergence of the internet during this period that led to an explosion of film reviews attacking the film's

Figure 16.1 Pamphlet on the creationist origins of *Jurassic Park*'s dinosaurs disseminated by the Southwest Radio Church.

position on evolution. The lack of gatekeepers for this new medium meant that anybody could disseminate their ideas online about the blasphemous science of *Jurassic Park*. A large number of anti-*Jurassic Park* webpages sprang up soon after the film's release, including one hosted by Probe Ministries (Bohlin, 1995). Fundamentalist Christian communities were not the only religious groups upset about the film's pro-evolution narrative. Some Orthodox Jews protested against the use of *Jurassic Park* promotional material on milk cartons in Israel, because they believed that 'dinosaurs symbolise a heresy of creation' (Goldman, 1993: 7).

Nowadays, while official censorship is no longer a threat, it is possible to indirectly censor a movie through means other than directly changing a script or banning a film. Movies can face a de facto ban if theatres are unwilling to show the film or if the film is unable to find distribution. This was the case for the 2009 film *Creation*, which was unable to initially find a distributor in the USA because its sympathetic portrait of Charles Darwin was considered to be 'too controversial' (Singh, 2009). According to Christian commentators this was not censorship; it was an example of market forces in action (Silvestru, 2009). From their perspective, it was not because of its subject matter that conservative Christian groups were keeping the film out of the USA. Rather, they believed that distributors had decided for financial reasons that the film would not be able to find an audience in a country in which only 39% of the population believed in the theory of evolution (Newport, 2009). In the end, *Creation* received a limited distribution through Newmarket Films, a company that specialised in distributing controversial films, including Mel Gibson's *The Passion of the Christ* (2004).

The same economic concerns about potential Catholic protests that fuelled the development of the PCA and the Legion of Decency back in the 1930s also drove studios to consult the Catholic Church during the production of two films in the late 2000s. *The Golden Compass* (2007) and *Angels and Demons* (2009) were both based on controversial books whose plots revolved around depictions of the Catholic Church as an organisation that actively obstructs scientific progress. In order to avoid Catholic boycotts of the film adaptations, the studios substantially reduced or removed any indication of an anti-science stance on the part of the Church. In a move reminiscent of the Legion of Decency, the studios also showed rough-print versions to Church officials while indicating that they might be willing to edit out any problematical elements (Pacatte, 2011). In spite of the studios' attempts to appease Catholic viewers, the films still ran into significant opposition and boycotts from Catholic organisations.

Although some films during this period feature contestation narratives about science and religion, filmmakers have also crafted a number of movies that function as reconciliation stories. Space exploration films such as *Contact* (1997), *Gravity* (2013) and *Interstellar* (2014) use a sense of wonder about the universe to introduce metaphysical ambiguities that can be understood as both scientifically and religiously inspired. Several recent films include scientist characters struggling with their faith in the face of scientific discoveries, such as *Knowing* (2009) and *Prometheus* (2012). Despite sympathetic portrayals of both science and faith, the Christian community's responses to these films were mixed. Christian commentators received *Knowing*'s message of benevolent extra-terrestrials rekindling a scientist's religious faith warmly (DeMar, 2009). *Interstellar*'s almost spiritual exploration of themes relating to love, death and sacrifice also resonated with many Christian reviewers (McCracken, 2014). On the other hand, while some Christians were pleased with the scientist's religiosity in *Prometheus*, most were disturbed by the notion of ancient alien creators in the film. Despite the earlier award for the similarly themed *2001*, the *Catholic News Service*'s review of *Prometheus* found that the plotline of alien-directed human evolution 'renders "*Prometheus*" extremely problematic for viewers of faith' (McCarthy, 2012).

Reconciliation narratives were meant to appease religious audiences who might have taken offence at these films' clear reverence for science. But some filmmakers have gone even further, by crafting science-heavy films that are directly aimed at courting religious audiences. The box office success of *The Passion of the Christ* provided a blueprint of how to use grassroots marketing to attract the religious right (Russell, 2013). Two films in the mid-2010s, *Noah* (2014) and *Exodus: Gods and Kings* (2014), used this blueprint to target religious audiences. But the directors of both films consciously attempted to frame traditional religious narratives as scientifically viable in order to also appeal to secular audiences (Bowman, 2016).

Ridley Scott looked to scientific rationales rather than miracles to explain the parting of the Red Sea and the ten plagues of Egypt in

Exodus (Vilkomerson, 2014), while Darren Aronofsky openly merged religion with science in *Noah* (Chattaway, 2014). Although they were adaptations of biblical stories, neither film managed to garner the approval of religious audiences. *Noah* proved to be problematic for religious audiences who rejected the science-based creation narrative as well as *Noah*'s obsessive focus on contemporaneous environmental concerns (Masters, 2014). In the cases of *Noah* and *Exodus*, religious audiences rejected science's intrusion into their stories of faith, while the scientific explanations were not enough to attract secular audiences to these biblical tales.

The recent proliferation of streaming services such as Netflix has meant that Christian films have become available to significantly larger audiences. Improved production values also mean that Christian films are often indistinguishable from major Hollywood movies. Many of the most successful mainstream Christian films have explored scientific and medical themes, including *October Baby* (2011), *God's Not Dead* (2014) and *Heaven Is For Real* (2014) (Macauley, 2015). Since 1968, religious organisations can no longer exert direct control over the scientific content in mainstream Hollywood movies. The current strategy for religious groups is to produce their own cinematic stories about science, and they have experienced a modicum of success in this outside their traditional Christian audience.

Conclusions

The created nature of movies makes them useful in understanding society's relationship with science because movies reveal the kinds of stories people want to tell about science. Filmmakers have made specific decisions to tell stories about science in particular ways. But our research demonstrates that decisions about scientific depictions in movies were not always left solely in the hands of filmmakers. Since the beginnings of cinema, religious groups in America have tried to influence the way that filmmakers used science to tell their cinematic stories, or they have tried to influence the way audiences interpret these stories about science. Religious organisations based their approach on simplistic assumptions about the nature of movies and the nature of communication. From their perspective, films told linear stories using a heightened visual realism that conveyed easily understandable narratives to a monolithic audience. From this simplistic viewpoint, cinema seemed to be a powerful force in determining our perceptions of the world. As such, they were concerned about the ways that movies portrayed science's role in society, science's status as a knowledge producer and science's relationship to the spiritual.

In cinema's early days, religious reformers believed that by controlling the content of scripts and distribution of finished films they could ensure that movies disseminated only morally or theologically appropriate messages about science. For many religious groups, censorship seemed to be a rational response to the dangers of cinema, especially at a time when activists were using the medium to promote scientific solutions to sexually based social issues such as VD, birth control and eugenics. Anticipating censure or boycotts forced filmmakers to make decisions about what science to include or remove, based on reasons that had nothing to do with artistic merit. In the case of the Hays Office, the PCA and the Catholic Legion of Decency, censorship decisions were founded on beliefs rooted in mid-twentieth-century American Christianity. These organisations' sense of moral certainty did not require their censors to understand the scientific topics upon which they were passing judgement, including evolution, psychiatry and atomic science.

When filmmakers were no longer under the threat of censorship, they could address more serious science-based topics, including environmental issues and biomedical ethics. This meant that religious groups had to change their tactics to address the scientific messages in films when direct censorship was no longer an option. Instead of preventing immoral messages, they decided to encourage studios by giving awards to films containing what they considered to be morally and theologically appropriate messages about the uses of science. Religious groups also began to provide their own movie reviews as a way to influence audiences' interpretations of scientific stories in films. These reviews allowed them to call attention to themes they found problematic. Reviews were also a means by which groups could celebrate narratives about science that they found inspirational, such as films promoting the spiritual nature of science. Mainstream filmmakers subsequently realised that they could achieve greater box office successes for their science-based films by incorporating scientific themes that appealed to Christian audiences. Ultimately, the Christian film industry decided that the easiest and, for their purposes, perhaps best way to control scientific themes in movies was by creating their own science-based films.

Acknowledgement

This work was supported by the Wellcome Trust (100618).

References

Black, G. (1996). *Hollywood Censored: Morality Codes, Catholics, and the Movies*. Cambridge: Cambridge University Press.

Bohlin, R. (1995). The worldview of *Jurassic Park*: A biblical Christian assessment. *Probe Ministries*. Retrieved 15 March 2016 from: www.probe.org/the-worldview-of-jurassic-park.

Bowman, D. (2016). The screen and the Cross: Christianity, Hollywood, and the culture wars. In S. Mintz, R. Roberts and D. Welky (eds), *Hollywood's America: Understanding History Through Film* (pp. 349–356). Chichester: John Wiley and Sons.

Brossard, D., Scheufele, D. A., Kim, E., and Lewenstein, B. V. (2009). Religiosity as a perceptual filter: Examining processes of opinion formation about nanotechnology. *Public Understanding of Science*, 18(5), 546–558.

Chattaway, P. (2014). Darren Aronofsky and Ari Handel on biblical accuracy and combining science and religion in *Noah*. *Patheos*, 26 March. Retrieved 17 March 2016 from: www.patheos.com/blogs/filmchat/2014/03/exclusive-darren-aronofsky-and-ari-handel-on-biblical-accuracy-and-combining-science-and-religion-in-noah.html.

Cobb, J. (1970). Ecological disaster and the Church. *Christian Century*, 7 October, 1185–1187.

Curran, C. E. (2008). *Catholic Moral Theology in the United States: A History*. Washington, DC: Georgetown University Press.

Dart, J. (1978). *Star Wars*: Religious impact in parable form. *Los Angeles Times*, 1 May, E11.

DeMar, G. (2009). Aliens as cosmic saviors. *American Vision*, 31 March. Retrieved 16 March 2016 from: www.americanvision.org/1210/aliens-as-cosmic-saviors.

Dickerson, J. (1993). Review of *Jurassic Park*. *Christian Spotlight on Entertainment*. Retrieved 15 March 2016 from: www.christiananswers.net/spotlight/movies/pre2000/i-jpark.html.

Evans, J. H., and Evans, M. S. (2008). Religion and science: Beyond the epistemological conflict narrative. *Annual Review of Sociology*, 34, 87–105.

Ferngren, G. B. (ed.) (2002). *Science and Religion: A Historical Introduction.* Baltimore, MD: Johns Hopkins University Press.

Gilbert, J. (1997). *Redeeming Culture: American Religion in an Age of Science.* Chicago: University of Chicago Press.

Gillis, C. (1999). *Roman Catholicism in America.* New York: Columbia University Press.

Goldman, A. (1993). Dinosaurs vs. the Bible. *New York Times*, 14 August, 7.

Grieveson, L. (2004). *Policing Cinema: Movies and Censorship in Early-Twentieth-Century America.* Berkley, CA: University of California Press.

Kavanaugh, J. F. (1971). *Hellstrom Chronicle* fascinates despite poor acting, editing. *St Louis Review* [news publication of the Roman Catholic Archdiocese of Saint Louis, Missouri]. 10 September, 12.

Kirby, D. A. (2011). *Lab Coats in Hollywood: Science, Scientists, and Cinema.* Cambridge, MA: MIT Press.

Kirby, D. A. (2014). Censoring science in 1930s and 1940s Hollywood cinema. In K. Grazier, D. Nelson, J. Paglia and S. Perkowitz (eds), *Hollywood Chemistry* (pp. 229–240). New York: Oxford University Press.

Lederer, S. E. (1998). Repellent subjects: Hollywood censorship and surgical images in the 1930s. *Literature and Medicine*, 17, 91–113.

Lederer, S. E., and Parascandola, J. (1998). Screening syphilis: *Dr. Ehrlich's Magic Bullet* meets the Public Health Service. *Journal of the History of Medicine and Allied Sciences*, 53(4), 345–370.

Leff, L. J., and Simmons, J. L. (2001). *The Dame in the Kimono: Hollywood Censorship and the Production Code.* Berkley, CA: University of California Press.

Linnell, G. (2006). 'Applauding the good and condemning the bad': The *Christian Herald* and varieties of Protestant response to Hollywood in the 1950s. *Journal of Religion and Popular Culture*, 12(1), doi: 10.3138/jrpc.12.1.004.

Macauley, W. R. (2015). Soul survivors: Faith, science, and redemption in Evangelical Christian films. Paper presented at the University of East Anglia, Faculty of Arts and Humanities seminar series, 26 October.

McCarthy, J. (2012). Review of *Prometheus*. *Catholic News Service*, 7 June. Retrieved 16 March 2016 from: www.catholicnews.com/services/englishnews/2012/prometheus.cfm.

McCracken, B. (2014). Review of *Interstellar*. *Christianity Today*, 6 November. Retrieved 16 March 2016 from: www.christianitytoday.com/ct/2014/november-web-only/interstellar.html.

Masters, K. (2014). Rough seas on *Noah*: Darren Aronofsky opens up on the biblical battle to woo Christians (and everyone else). *Hollywood Reporter*,

21 February. Retrieved 20 March 2016 from: www.hollywoodreporter.com/news/rough-seas-noah-darren-aronofsky-679315.
NCOMP (1974). *The Exorcist. Catholic Film Newsletter*, 39(1), 2.
Neale, S. (2005). 'The last good time we ever had?': Revising the Hollywood renaissance. In L. R. Williams and M. Hammond (eds), *Contemporary American Cinema* (pp. 90–108). Maidenhead: McGraw-Hill.
Newport, F. (2009). On Darwin's birthday, only 4 in 10 believe in evolution. *GALLUP*, 11 February. Retrieved 17 March 2016 from: www.gallup.com/poll/114544/Darwin-Birthday-Believe-Evolution.aspx.
Nisbet, M., and Scheufele, D. A. (2009). What's next for science communication? Promising directions and lingering distractions. *American Journal of Botany*, 96(10), 1767–1778.
Olasky, M. N. (1985). The failure of movie industry public relations, 1921–1934. *Journal of Popular Film and Television*, 12(4), 163–170.
Ostherr, K. (2013). *Medical Visions: Producing the Patient Through Film, Television, and Imaging Technologies*. Oxford: Oxford University Press.
Pacatte, R. (2011). Holy moments: Faith and values at the movies. *New Theology Review*, 24(3), 6–16.
Parry, M. (2013). *Broadcasting Birth Control: Mass Media and Family Planning*. Rutgers, NJ: Rutgers University Press.
Pernick, M. (1996). *The Black Stork: Eugenics and the Death of 'Defective' Babies in American Medicine and Motion Pictures since 1915*. Oxford: Oxford University Press.
Pernick, M. (2007). More than illustrations: Early twentieth-century health films as contributors to the histories of medicine and of motion pictures. In L. J. Reagan, N. Tomes and P. A. Treichler (eds), *Medicine's Moving Pictures: Medicine, Health, and Bodies in American Film and Television* (pp. 19–35). Rochester, NY: University of Rochester Press.
Romanowski, W. (2012). *Reforming Hollywood: How American Protestants Fought for Freedom at the Movies*. Oxford: Oxford University Press.
Russell, J. (2013). In Hollywood, but not of Hollywood. In G. King, C. Molloy and Y. Tzioumakis (eds), *American Independent Cinema: Indie, Indiewood and Beyond* (pp. 185–197). London: Routledge.
Schaefer, E. (1999). *Bold! Daring! Shocking! True! A History of Exploitation Films, 1919–1959*. Durham, NC: Duke University Press.
Shelley, M. (1998). *Frankenstein, or, the Modern Prometheus: The 1818 Text*. Oxford: Oxford University Press.
Scheufele, D. A., Corley, E. A., Shih, T.-J., Dalrymple, K. A., and Ho, S. S. (2009). Religious beliefs and public attitudes toward nanotechnology in Europe and the United States. *Nature Nanotechnology*, 4(2), 91–94.

Silvestru, E. (2009). Rejecting *Creation* the movie: A business decision, film distributors not imposing their ideology. *Creation.com*, 10 December. Retrieved 17 March 2016 from: http://creation.com/rejecting-creation-the-movie.

Singh, A. (2009). Charles Darwin film 'too controversial for religious America'. *Telegraph*, 11 September. Retrieved 17 March 2016 from: www.telegraph.co.uk/news/worldnews/northamerica/usa/6173399/Charles-Darwin-film-too-controversial-for-religious-America.html.

Siska, W. (1977). A break of fresh fantasy: Review of *Star Wars*. *Christian Century*, 20 July, 666–668.

Stevenson, R. L. (2000 [1886]). *The Strange Case of Dr Jekyll and Mr Hyde*. New York: Scribner (originally published 1886).

Stoker, B. (2003 [1897]). *Dracula*. London: Penguin (originally published 1897).

Toumey, C. (2011). Seven religious reactions to nanotechnology. *Nanoethics*, 5, 251–267.

Vilkomerson, S. (2014). How Ridley Scott looked to science – not miracles – to part the Red Sea in *Exodus: Gods and Kings*. *Entertainment Weekly*, 23 October. Retrieved 7 March 2016 from: www.ew.com/article/2014/10/23/ridley-scott-red-sea-exodus.

Walsh, F. (1996). *Sin and Censorship: The Catholic Church and the Motion Picture Industry*. New Haven, CT: Yale University Press.

Wells, H. G. (1996 [1896]). *The Island of Dr. Moreau*. New York: Modern Library (originally published 1896).

Wyatt, J. (2000). The stigma of X: Adult cinema and the institution of the MPAA ratings system. In M. Bernstein (ed.), *Controlling Hollywood: Censorship and Regulation in the Studio Era* (pp. 238–264). London: Athlone Press.

Films referenced

2001: A Space Odyssey (1968). Dir. S. Kubrick. MGM.
The Amazing Dr Clitterhouse (1938). Dir. A. Litvak. Warner Bros.
The Andromeda Strain (1971). Dir. R. Wise. Universal Pictures.
Angels and Demons (2009). Dir. R. Howard. Columbia Pictures.
Back to the Future (1985). Dir. R. Zemeckis. Universal Pictures.
Before I Hang (1940). Dir. N. Grinde. Columbia Pictures.
The Beginning or the End (1947). Dir. N. Taurog. MGM.
Birth Control (1917). Dir. M. Sanger. B. S. Moss Motion Picture Corporation.
Captive Wild Woman (1943). Dir. Edward Dmytryk. Universal Pictures.
A Clockwork Orange (1971). Dir. S. Kubrick. Warner Bros.
Close Encounters of the Third Kind (1977). Dir. S. Spielberg. Columbia Pictures.
Contact (1997). Dir. R. Zemeckis. Warner Bros.
Creation (2009). Dir. J. Amiel. Recorded Picture Company.

Damaged Goods (1914). Dir. T. Ricketts. American Film Manufacturing Company.
Damaged Goods (1937). Dir. P. Stone. Criterion Pictures Corp.
Dr Ehrlich's Magic Bullet (1940). Dir. W. Dieterle. Warner Bros.
Dr Renault's Secret (1942). Dr. Harry Lachman. Twentieth Century Fox.
The End of the Road (1918). Dir. E. H. Griffith. American Social Hygiene Association.
E.T. the Extra-Terrestrial (1982). Dir. S. Spielberg. Universal Pictures.
Exodus: Gods and Kings (2014). Dir. R. Scott. Scott Free Productions.
The Exorcist (1973). Dir. William Friedkin. Warner Bros.
Fit to Fight (1918). Dir. E. H. Griffin. American Social Hygiene Association.
Frankenstein (1931). Dir. J. Whale. Universal Pictures.
Freud (1962). Dir. J. Huston. Universal International Pictures.
God's Not Dead (2014). Dir. H. Cronk. Pure Flix Productions.
The Golden Compass (2007). Dir. C. Weitz. Newline Cinema.
Gravity (2013). Dir. A. Cuarón. Warner Bros.
Heaven Is For Real (2014). Dir. R. Wallace. TriStar Pictures.
Heredity (1915). Dir. W. Humphrey. Vitagraph Company of America.
The House Without Children (1919). Dir. S. Brodsky. Argus Motion Picture Company.
Inherit the Wind (1960). Dir. S. Kramer. Lomitas Productions Inc.
The Invisible Man (1933). Dir. J. Whale. Universal Pictures.
Interstellar (2014). Dir. C. Nolan. Paramount Pictures.
Island of Lost Souls (1932). Dir. E. C. Kenton. Paramount Pictures.
I've Lived Before (1956). Dir. R. Bartlett. Universal International Pictures.
Jurassic Park (1993). Dir. S. Spielberg. Universal Pictures.
Knowing (2009). Dir. A. Proyas. Summit Entertainment.
Madame Curie (1943). Dir. M. Lery and A. Lewin. Metro-Goldwyn-Meyer.
The Man with Two Lives (1942). Dir. P. Rosen. A. W. Hackel Productions.
Men in White (1934). Dir. R. Boleslavsky. Metro-Goldwyn-Meyer.
Noah (2014). Dir. D. Aronofsky. Paramount Pictures.
October Baby (2011). Dir. A. Erwin and J. Erwin. Gravitas/Provident Films.
The Omega Man (1971). Dir. B. Sagal. Walter Selzer Productions.
The Passion of the Christ (2004). Dir. M. Gibson. Icon Productions.
Planet of the Apes (1968). Dir. F. J. Schaffner. APJAC Productions.
Prometheus (2012). Dir. R. Scott. Twentieth Century Fox Film Corporation.
Red Planet Mars (1952). Dir. H. Horner. Melaby Pictures.
The Regeneration of Margaret (1916). Dir. Anon. Essanay Film Manufacturing Company.
The Scarlet Trail (1918). Dir. J. S. Lawrence. G. and L. Features Inc.
Scott of the Antarctic (1948). Dir. C. Frend. Ealing Studios.

Silent Running (1972). Dir. D. Trumbull. Universal Pictures.
Sister Kenny (1946). Dir. D. Nichols. RKO Radio Pictures.
Spellbound (1945). Dir. A. Hitchcock. Selznick International Pictures.
The Spreading Evil (1918). Dir. J. Keane. James Keane Feature Photo-play Productions.
Soylent Green (1973). Dir. R. Fleisher. Metro-Goldwyn-Meyer.
Star Wars (1977). Dir. G. Lucas. Twentieth Century Fox.
The Terminal Man (1974). Dir. M. Hodges. Warner Bros.
The War Game (1965). Dir. P. Watkins. British Broadcasting Corporation.
Zardoz (1974). Dir. J. Boorman. John Boorman Productions.

Afterword: monstrous markets – neo-liberalism, populism and the demise of the public university

John Holmwood, Jan Balon

There is a crisis in the idea of the university. It has emerged from the application of neo-liberal policies which have reduced the public values of the university to instrumental purposes. This poses a considerable threat to liberal education (Brown, 2015, Collini, 2012; Ginsberg, 2011; Holmwood, 2011; Nussbaum, 2010). In the UK, government ministers and policy advisers seek a 'cultural' change directing academic research and student recruitment towards the market and in service of a global knowledge economy. There are few dissenting voices among those with institutional responsibility for the academy – namely, its vice-chancellors and senior representatives. Vice-chancellors have not spoken out to protect the university's wider public values, and few learned societies have either. Senior university personnel have mostly been interested in maintaining funding, especially in the context of the politics of austerity after the financial crisis of 2008 (Smith, 2011). Learned societies and research councils have had similar concerns about funding, and have been concerned to establish the utility of research, especially in the context of the impact agenda. Dissent comes mainly from some individual academics and from students. The latter have experienced a dramatic rise in their costs alongside diminishing labour market opportunities, notwithstanding an emphasis on their private investment in human capital as a justification of the reduction in the public funding of undergraduate education.

In part, this quiescence itself derives from a mode of governance specific to neo-liberalism which operates through the co-production of policy objectives. This involves consultation with those affected by proposed policies and with interests in the outcome, generally called the 'stakeholders'. Consultation might appear to be an *evidence-based* process with consensus as its aim, but, in truth, interests are frequently not reconcilable and what the parties put forward is *interest-based* evidence. In this context, government acts as mediator of such evidence, which it collates and selects according to its own policy objectives while managing alternative views. At the same time, stakeholders also lobby government independently of the consultation process. In this way, consultation operates in the interest of the most powerful stakeholders and requires wider publics (who might bear the consequences of the policies) to be represented by a 'stakeholder' or accept the fiction that it is the government itself that represents their interests (for example, as 'taxpayers', or as the guardian of the interests of students as 'consumers').

In the case of recent university reforms in the UK – which have shifted from direct public funding of undergraduate higher education to what is primarily fee-based funding via a system of publicly supported student loans – the government retains the ability to determine the revenue received by universities and so can maintain compliance from vice-chancellors and representative bodies, while opening the sector to for-profit providers and allowing the title of 'university' to single-subject, teaching-only entities. In this way, despite the UK government proposing the most fundamental changes to higher education, this has occurred with little active debate or challenge to the underlying market logic that guides those changes. The university is under threat, but all universities are busy 'co-producing' these changes and have passed their voices into the dominant neo-liberal discourse.

The problem of populism

The place of the university within public culture is not separate from the fate of public culture itself. To some extent, the crisis in the idea of the university reflects a crisis of public culture, one that has become most evident in the rise of the 'far right' and 'populism'. The UK referendum vote to leave the European Union ('Brexit') and the election of Donald Trump as President of the United States are each widely understood as involving a 'populist' rejection of 'elites' and 'economic globalisation'. It is significant that each has taken place in a country where neo-liberal public policy has been paramount. However, the rise of authoritarian populist regimes elsewhere (for example, in Turkey and India) and of far-right political parties having increasing political influence (for example, the National Front in France, Sweden Democrats, and the Freedom Party of Austria) indicates that 'populism' is a more general issue. It has also been reported that Chinese President Xi Jinping has called for intensified ideological control over universities (Philips, 2016), including, presumably, the sixty-four 'branch' operations of transnational higher-education institutions currently operating in China, of which Nottingham University Ningbo is one (He, 2016).

Oxford Dictionaries (2016) has marked this new political mood by announcing online that 'post-truth' was its 'word of the year', 'denoting circumstances in which objective facts are less influential in shaping public opinion than appeals to emotion and personal belief'. This followed a statement in a television interview by the then British Minister for Justice and Vote Leave campaigner, Michael Gove, that he thought that 'the British Public have had enough of experts'. Where does this leave the university? The trade magazine of the profession in the UK, *Times Higher Education*, has suggested that the very intellectual character of universities is anti-populist and, thus, that they contribute to a polarisation of politics rather than being able to moderate the effects of such a polarisation (Morgan, 2016). Indeed, the report commented that 'a *THE* poll before the vote found that 88.5 per cent of university staff intended to vote Remain and 9.5 per cent Leave. That was just a shade out of line with the actual UK result, which saw 48.1 per cent vote Remain and 51.9 per cent vote Leave' (Morgan, 2016: n.p.).

The article also suggested that universities might 'reposition themselves as the voices of moderation. In other words, as the populists head off to extremes, some believe that US universities should move more towards the centre politically – or rightwards from where they currently are – in an attempt to "depolarise" their relationship to wider society' (Morgan, 2016). If universities, including those in the UK, do not do so, the article warned, 'they will clearly need to tread very carefully lest they portray themselves as part of the global elite resented by populist supporters. Otherwise, they will only intensify the risks to their funding, their culture and their educative missions' (Morgan, 2016).

However, this is a poor understanding of populism and the problems it poses for universities. As Müller (2016) has recently argued, it is the very nature of populism to represent itself as speaking for the 'people', with pluralism seen as 'bad faith'. There can be no 'de-polarisation' where populism is identified as one of the polar positions, since it admits no compromise. In this way, according to Müller, populism is both a product of representative democracy and, at the same time, a denial of democracy since it depends on 'othering' those it opposes as reflecting the interests of deracinated cosmopolitan 'elites' and without a legitimate voice in 'democratic' debate.[1]

In both the USA and the UK, 'populism' is also a form of 'nativism' manifest in calls to 'take back our country', with its hostility both to external powers that might limit the scope of action and to those within the nation who are not seen as properly part of it. In both cases, those who are not part of the 'we' are racialised minorities, immigrants and what Enoch Powell in the UK context once chillingly called the 'immigrant descended'. In the UK, Dame Louise Casey (2016), in a recent review into 'Opportunity and Integration', has called for migrants to 'swear an oath of allegiance' on arrival into the UK, while schools, under the Prevent agenda, are obliged to teach 'British values'.

Danielle Allen (2004) has made a similar argument to Müller's about the problematic idea of popular sovereignty that is frequently represented as a republican ideal. For example, incorporated in the US Declaration of Independence and reproduced daily in US schools is the Pledge of Allegiance to 'one Nation indivisible with liberty and justice for all'. The idea of 'one Nation indivisible' implicitly passes all voices into one, but what would happen, Allen asks, were we to propose instead an allegiance to the '*whole* Nation indivisible'? The *whole* nation would be understood as a nation of parts – that is, as differentiated – and an obligation towards indivisibility would be an obligation towards difference and its recognition.

What does this mean for the nature and culture of the university; that is, for the role of higher education in the public life of a

---

1 As newly appointed post-Brexit UK Prime Minister Theresa May put it (buying into the populist mood), 'if you are a citizen of the world, you are a citizen of nowhere' (May, 2016).

nation? We shall suggest that one aspect of this role must be the facilitation of inclusive democratic public debate. Nor could this be understood simply as providing the knowledge that might form the evidence base for public policy where the latter is directed at policymakers rather than at wider publics, as argued by the Campaign for Social Science (2015). Morgan suggests that there is a risk that universities will be perceived as aligned with a global elite and that this might cost them their funding. However, part of the problem is that universities have put their cultural and educative mission at risk precisely because of their concern with funding, while ignoring how the conditions of that funding have been tied to a change in their mission.

In this afterword, we will draw on the work of John Dewey, especially his *The Public and Its Problems* (1927), to suggest that populism is a problem of 'publics' and the institutions in the public sphere that support them. In brief, we shall argue that neo-liberalism represents an attempt to replace publics with markets, a process that is facilitated by the privatisation of public institutions, including that of the university itself. This 'hollowing out' of the public sphere is precisely what creates the space for populism. Neo-liberalism requires a strong interventionist state on behalf of markets, but it also requires democratic legitimation. This makes populism an ideology of justification (Boltanski and Thévenot, 2006) supplementary to that of the market – one that is mobilised against the public sphere, which has come to be characterised as dominated by liberal elites, notwithstanding that the promotion of the market itself operates to widen inequalities.

The neo-liberal knowledge regime

Changing inequalities are particularly significant for universities, but this is something that has been relatively neglected. As Clark Kerr, architect of the California 'Master Plan' (initiated at the same time as the Robbins reforms in the UK in the 1960s), argued, the rise of mass higher education and public funding would make the university increasingly subject to political scrutiny (Committee on Higher Education, 1963; Kerr, 2001 [1963]). The expansion of public higher education was not a simple extension of arguments that had justified public secondary education and its compulsory nature. The latter was universal in character and, therefore, could be represented as a 'social right' that secured a public benefit; namely, a common education for citizens, a benefit recognised even by Milton Friedman (1962). Participation in public higher education was not intended to be universal, merely to be expanded (in the UK case, closer to the level already attained in the US). In this context, there was potentially the issue that higher education secured a private benefit for those who graduated from it, when compared with those that did not.

At the same time, no matter how much participation might be widened, it would be likely to attract proportionally more of its participants from socially advantaged backgrounds. However, at the time, there was a general expectation of a shift from an industrial to a post-industrial, knowledge-based, economy, where there would be increased demand for educated labour and a general 'adaptive upgrading' of all jobs. Indeed, this was evident in the way in which a secular trend in the reduction of inequalities was regarded as 'institutionalised' across most Western societies, even if the level of inequalities was significantly greater in some (the US) than in others (Sweden, or the UK up until the 1980s). In effect, this was endorsed as 'fact' by Kuznets (1953) and his 'curve' demonstrating how declining income inequality emerged alongside economic growth.

Public spending on higher education, then, could be justified in terms of its wider benefits; even if an individual's educational attainments and preferences did not take him or her to university, there would be a benefit from the greater integration of higher education and the economy. The economic growth to which expanded higher education and research would contribute was understood to be *inclusive*, associated with what was perceived to be a secular decline in inequality. This also included changes in what might be regarded as the 'status order' of employment relations, as the terms of the labour contract became more similar across manual and non-manual work, and rights previously enjoyed by non-manual workers were extended to all employees. This idea of inclusive economic growth was integral to the idea of an emerging 'knowledge society' – as distinct from a 'knowledge economy' – and, in the telling phrase used by Clark Kerr (2001 [1963]), what had emerged was a 'multiversity' meeting multiple functions – direct economic functions, certainly, but also wider social functions, including amelioration and democratisation. In other words, higher education was part of a wider 'moral economy' underpinned by social rights (Holmwood and Bhambra, 2012).

It is precisely this 'moral economy' that is called into question by neo-liberalism, and not simply in terms of seeking to deny the existence of social rights. Wider neo-liberal policies have given rise to widening inequalities and reductions in taxation, especially progressive taxation. Moreover, the deregulation of labour markets has created new forms of labour contract and a new polarisation between 'good' and 'bad' jobs (Brown et al., 2011; Kalleberg, 2011). The function of higher education to support economic growth remains, but inclusive economic growth is no longer a government objective. In this context, government policies to reduce taxation put pressure on university funding, while widening inequality increased calls for the beneficiaries of higher education to pay. In the UK this was first introduced as a fee contribution by students alongside direct public funding in 1999, following the Dearing Review of 1997, but became wholly fee based for arts, humanities and social sciences in 2010 following the financial crisis of 2008 and cuts to government spending under the mantra of austerity.

The irony is that just as the argument that it is right that students should pay fees because they are private beneficiaries was being put forward, the opposite argument was made with regard to research. The UK Government put forward an 'impact' agenda, where all publicly funded research should be undertaken with specific 'beneficiaries' in mind. Here, the argument was that publicly funded research should show a direct benefit, but the beneficiary *should not pay*. In part, the purpose of the impact agenda was to shorten the time from 'idea to income' or the research-development cycle.

It might seem that this was a simple continuation of the perceived function of research for economic growth that was described by Kerr (2001 [1963]). However, the context is significantly different. First, as we have suggested, economic growth that is publicly funded is no longer inclusive in its benefits (see also Organisation for Economic Co-operation and Development (OECD), 2015). Second, the neo-liberal policies directed towards wider corporate governance have emphasised share-holder value, which has made companies more concerned with profits in the short term. In consequence, private investment in research and development has fallen, such that the UK has moved from having one of the largest investments (as a proportion of GDP) among OECD countries in the 1960s to now having one of the smallest (Jones, 2013). Finally, the emphasis on delivering the benefits of research over a shorter time has altered the balance between privately funded and publicly funded research within the research–development ecosystem, and undermined the longer-term and more fundamental benefits that publicly funded research can achieve (see Mazzucato, 2011).

Of course, the UK Government's impact agenda is wider than simple commercial benefits, however pronounced the latter are within it. It also includes impact on public policy and other aspects of social well-being. What is common across commercial and non-commercial impacts, however, is that impact has to be demonstrated with specific beneficiaries and that the strong recommendation of research councils is that this be done through 'co-production' of the research with them (that is, including likely beneficiaries at all stages of the research, including that of its design).

'Co-production' as a term derives from the work of Gibbons and his colleagues (1994), involving a distinction between 'mode 1' knowledge directed at academic audiences and 'mode 2' knowledge directed at non-academic audiences (see also chapter 6). The latter is frequently interdisciplinary applied-problem-solving knowledge, and the idea of co-production is used to capture the 'larger process in which discovery application and use are closely integrated' (Gibbons et al., 1994: 46). Gibbons and his colleagues did not anticipate that mode 2 knowledge would supplant mode 1 knowledge, but it is clear that the impact agenda promotes mode 2 knowledge.[2] Nor did they consider the wider environment in which co-production took place; that is, from the perspective of its beneficiaries.

It is clear that the beneficiaries are commercial organisations, government bodies (at national or local levels) or civil society actors (non-governmental organisations (NGOs) and charities, etc.). We have already suggested that neo-liberal public policy has the effect of hollowing out the public sphere. This takes place in two ways. The first

---

2 The argument for mode 2 knowledge has been extended by Etzkowitz (2008), and the idea of the 'triple helix' of interdependencies between government, industry and university.

is by the direct privatisation of public bodies, the second by recommending that charitable bodies and NGOs be involved in the provision of services. The latter also includes charities and other voluntary associations operating together with for-profit organisations. The emphasis on the co-production of research is part of a wider neo-liberal project that includes the binding of the beneficiaries into government policy by the fact that they are frequently dependent on government for their own funding.[3]

Tying civil-society organisations to government objectives involves a deformation of the public sphere that constitutes the context for the rise of populism. It is something in which universities are directly implicated. For example, the Academy of Social Sciences drew up a report under the auspices of its Campaign for Social Science – significantly, entitled *The Business of People* – to campaign for public funding of social science, prior to the 2015 election. This was the election that included a Conservative manifesto commitment to a referendum on membership of the European Union. The report was preceded by the widespread news coverage of Thomas Piketty's *Capital in the Twenty-First Century* (2014) with its depiction of widening social inequality. Concerns about inequality were also raised in OECD reports.

---

3 This became particularly evident following new Cabinet Office rules to prevent bodies in receipt of government funding from engaging in lobbying. This followed intense lobbying from a neo-liberal think tank, the Institute for Economic Affairs, against what it called 'sock puppets' (Snowdon, 2012). The action was directed against charities like Save the Children, Action on Smoking and Health, and Alcohol Concern. Universities were alarmed that they might be included and that the proposals were antithetical to the impact agenda. Universities were subsequently declared exempt and the proposals watered down. However, the implications for civil-society organisations and the vulnerability of their funding should they be perceived to be too politically active in pursuing their remit is clear. This is explicitly recognised by a report for the National Coalition of Independent Action: 'The force of entering the welfare market, increasingly as bid candy, has had disastrous consequences for voluntary services and their ability to respond to community needs. The capitulation by many in the voluntary sector, including its national and local leadership bodies, to these government agendas has done much damage to the ability of voluntary organisations to work with and represent the interests of individuals and communities under pressure. Privatisation and co-option into the market is driving down the conditions of staff working in voluntary services, diminishing their role in advocacy and jeopardising the safety of people using such services' (Waterson, 2014: 2).

Yet, structured social inequality is not mentioned at all in the report, nor is race and ethnicity, nor any other research on social structure. These profoundly affect the circumstances of people's lives, yet all the report has to say about them is of their derived consequences in terms of people's attitudes and behaviours and how those may be a problem for policymakers and practitioners in attaining their objectives. The report is overwhelmingly instrumental and designed to appeal to the 'Treasury, ministers, MPs and policy makers' (Campaign for Social Science, 2015: 'Foreword'). Its focus on policymakers and practitioners is unremitting: 'Advancing and applying science depends on profits, policies, markets, organisations and attitudes' (2015: 'Executive summary'). The attitudes of the public, on the other hand, are presented as potential obstacles to policy objectives. For example, it argues that 'study of public values and attitudes is vital, too, especially when innovation prompts uncertainties and concerns, as with genetically modified crops or shale gas extraction' (2015: 6). And it warns that 'without a better grasp of people, technological advances may be frustrated, or blocked, and fail to realise their potential' (2015: 5).

In short, a report on the value of the social sciences produced in the context of a general election made no reference at all to problems of inequality and no reference to their contribution to the facilitation of democratic debate. Instead, it was directed entirely at what populist rhetoric described as the 'political establishment' and its 'experts'.

The problem of democratic knowledge

As we suggested earlier, one of the problems in current understandings of the democratic significance of the university is that its role as an institution in the public sphere is weakly expressed (see also Holmwood, 2016b). In addition, government is allowed to stand as representing the 'public' and, in consequence, its policies constitute a working definition of the public interest. It is precisely these understandings that have been exposed by the rise of populism and are in need of reformulation.

We have also suggested that an alternative formulation can be found in the work of John Dewey, and especially his book on *The*

Public and Its Problems (1927). Significantly, for our purposes, the book was written in a similar context of an intense debate on the nature of the relation between expertise and democracy. Dewey was responding to the argument of Walter Lippmann (1925) that increased social complexity undermines the possibility of democracy being able to approximate the forms endorsed by standard liberal accounts of representative democracy. The public, for Lippmann, was increasingly ill equipped to make the sort of judgements attributed to them within democratic theory.

In consequence, he argues that the public is a 'phantom category' (that is, something that functions only in theories of democracy and has little real substance). For Lippmann, what Dewey came to call the 'eclipse of the public' is a necessary consequence of the complexity of modern societies that increasingly requires organised expertise of various kinds. In consequence, 'expert opinion' would replace 'public opinion' and democracy would necessarily be attenuated. Lippmann anticipated that expert opinion would operate in conjunction with the state and economic corporations and, in effect, would be 'co-produced' by them. However, it is significant that Lippmann also prefigured what would become another part of the neo-liberal solution; namely, the shift of decisions from the political sphere to the economic sphere, or from the state to the market.[4]

Dewey noted that the 'eclipse of the public' is prefigured in the very idea of the market economy, in which decisions by (consumer) sovereign individuals are perceived to be efficiently aggregated through impersonal market exchanges. This is held to be in contrast to their inefficient aggregation by collective political decision making through the agency of the state. In other words, according to Dewey, the idea of a political realm in which the public expresses its democratic will is already severely compromised by the liberal distrust of 'group', or collective, actions, and the idea that it is only the market that can properly express the general interest.

Dewey proposed to rescue the public from its eclipse by market and expert opinion alike by a radical refocusing of political philosophy,

---

4 He was a participant in the Colloque Walter Lippmann, which met in 1938 and was named in his honour. It was the first to coin the term 'neo-liberalism' for its position.

not as a *theory of the state* and its forms, but as a *theory of the public* and of the relation of institutional forms to the public, with the university as one crucial institutional form. He did so through an account of the 'social self', which he contrasted with the 'liberal self', as expressed in economics and political theory (in this way, also indicating the normative assumptions in the liberal idea of instrumental knowledge).

Dewey began from the argument that the individual is necessarily a social being involved in 'associative life', and that this is true of what are conventionally regarded as private actions as well as of public actions.[5] For Dewey, individuals form associations, but they are also formed by associations. At the same time, the multiplicity of associations and their interconnected actions have consequences. In all of this, Dewey's idea of a 'public', and of the several natures of 'publics', is crucial. It contains a strong idea of democracy associated with participation and dialogue, but does not deny that there will be functionally differentiated publics, whose articulation will be at issue. The key to his definition of a public is contained in the idea of action in the world having effects and consequences that are ramified and impact upon others who are not the initiators of the action. Essentially, all action is associative action, but a public is brought into being in consequence of being indirectly and seriously affected by those actions of others. His analysis of the problem of modern democracy, then, was concerned with the imbalance in the development of associations and the proliferation of problems in areas where the public cannot properly defend itself.

This immediately raises the issue of the state as the representative of the 'public'. It is the point at which Dewey shifted gear to argue that the wider idea of a public can achieve a level of generality that requires organisation and personnel to express it. This is the idea of a state understood as a set of public authorities. Thus, Dewey proposed that 'the lasting, extensive and serious consequences of associated activity bring into existence a public. In itself it is unorganized and

---

5 It is precisely this that Dewey suggested allows the understanding of the changing definition of the boundaries of what are conventionally regarded as private and public. The conventional definition of the 'private' is that of associated life that does not impinge with wider consequences upon others.

formless. By means of officials and their special powers it becomes a state. A public articulated and operating through representative officers is the state; there is no state without a government, but also there is none without the public' (1927: 67).

Dewey by no means suggested that these developments mean that a state necessarily will act in the public interest – power can be accrued, authority exercised despotically, and, indeed, the personnel of government can act on their own private or other special interests. The fundamental point, however, is that the state takes its meaning from the idea of a public and its interests, and that this is conceived as a dynamic thing. This means that, for Dewey, not only associations external to the state, but the state itself and its modes of organisation, are subject to change and revision in the light of other changes in the development of associative life. In other words, although the state exists in relation to the problems of associative, social life that create a public, its own forms and modes of organisation may come to constitute a problem for the expression of that public, although, paradoxically, that is its *raison d'être*.

Dewey had as his target two pathologies. The first *sets the state against the public* and is attributed to liberal individualism and its argument for the minimum state. The second is attributed to the conditions of modern corporate capitalism, in which there appears to be an '*eclipse of the public*' brought about by the dominance of corporate interests over the state. Dewey argued that the first undermines the individual as surely as it seeks to set the individual free. This is because the ruling idea of liberalism is that of the individual free of associations, which is linked with the idea of the 'naturalness' of economic laws (embodied in market exchanges). It is precisely the ideology of liberal individualism, according to Dewey, that suggests that the market can replace the state as the regulator of social life, but leaves the individual vulnerable to the outcomes of the market.

However, according to Dewey, this doctrine emerged just as the idea of an 'individual' free of associations was being rendered untenable by the very developments of corporate capitalism with which it was linked. Thus, Dewey said that ' "the individual", about which the new philosophy centred itself, was in process of complete submergence in fact at the very time in which he was being elevated on high in theory'

(1927: 96). The ideology which operates in the name of the individual, then, serves to undermine the very protection of the individual from egoistic, corporate associations that are themselves the very antithesis of the doctrine being espoused.

For Dewey, what is necessary for the proper expression of the public and for democracy is a 'Great Community'. Without it, there would be nothing more than state-supported corporate interests, together with partial and ad hoc responses. In contrast, Dewey wrote of democracy in the 'Great Community' that, 'from the standpoint of the individual, it consists in having a responsible share according to capacity in forming and directing the activities of the groups to which one belongs and in participating according to need in the values which the groups sustain. From the standpoint of the groups, it demands liberation of the potentialities of members of a group in harmony with the interests and goods which are in common. Since every individual is a member of many groups this specification cannot be fulfilled except when different groups interact flexibly and fully in connections with other groups' (1927: 147).[6]

Reclaiming the public university

Dewey was also writing at the birth of the 'multiversity'. Knowledge production and professional services were coming increasingly to be university based, and, at the same time, the university was becoming increasingly involved in the corporate economy with the commodification of research. As I have argued, these are all aspects of our current impact agenda. Yet, Dewey wished to argue that the university has a role for democracy and in facilitating the Great Community. The final issue I want to address is whether the complexity attributed to contemporary society and the problems it poses for a democratic public can be answered by the role of 'experts'.

Quite apart from the undemocratic implications of Lippmann's claim that experts can represent publics, Dewey also challenged it on sociological grounds. Co-production takes the structure of associations

---

6 The transformation of university decision making from collegial to hierarchical, managerial modes of organisation is significant in the context of this quotation (see Holmwood, 2016a).

as given, when the issue of publics is always how they are to be brought into a responsible share in the direction of activities. 'Co-production' is necessarily based upon forms of inclusion and exclusion. Finally, while the operation of economic interests can be unseen, precisely because of the formal separation of economic and political institutions typical of modern capitalism, the application of expert knowledge must necessarily take place in front of the public.

While the argument about the role of experts depends upon the idea that the public is unable to judge complex matters, it remains the case that it will be able to judge the pretensions of experts. Moreover, it is likely to be vulnerable to populist mobilisations by the very interests that expert opinion is being called upon to moderate. Thus, Dewey wrote that 'rule by an economic class may be disguised from the masses; rule by experts could not be covered up. It could only be made to work only if the intellectuals became the willing tools of big economic interests. Otherwise they would have to ally themselves with the masses, and that implies, once more, a share in government by the latter' (1927: 206). As soon as 'expertise' is defined in terms of the instrumentalisation of knowledge, there arises the problem that it is aligned with interests and, thereby, a problem of trust.

Dewey's concern with the problem of experts and their relation to wider publics speaks directly to our own circumstances. As expertise is increasingly co-produced, what seems to be attenuated is the role of the wider public. In a context where risks of concentrated activities – whether of nuclear power production or carbon-hungry economic profit seeking, to give just two examples – are also seen to be widely (indeed, globally) distributed, those that are affected are displaced from participation in decisions about them. At the same time, the nature of democracy is that wider public opinions can be made to count in elections and are subject to populist influence by advertising and by mass media, precisely as Dewey set out.

For Dewey, however, the significance of expert knowledge is how it can facilitate public debate, not government and corporate decision making independent of the participation of the wider public. The character of expert knowledge increasingly embedded within corporations and government serves to delegitimate expertise precisely by these forms of associations. It is necessarily part of the 'eclipse of the public'. As Dewey put it, 'the essential need … is the improvement of the methods and conditions of debate, discussion and persuasion. That is the problem of the public' (1927: 208).

If the improvement of debate, discussion and persuasion is the problem, then the university is necessarily part of the answer. But it is only part of the answer if it is at the service of the public. A university at the service of the public, in Dewey's sense, is a university that should properly be regarded as a public university. This would not be the only function of a university, but it is a necessary function and it is one that would place social justice at the heart of community engagement. Anything less and the university is just another private corporation in which a corporate economy has become a corporate society. The neo-liberal university would finally have given up any pretension to a social mission other than being at service to whoever paid.[7]

References

Allen, D. S. (2004). Invisible citizens: On exclusion and domination in Ralph Ellison and Hannah Arendt. In M. Williams and S. Macedo (eds), *Nomos XLVI: Political Exclusion and Domination* (pp. 29–76). New York: New York University Press.

Boltanski, L., and Thévenot, L. (2006). *On Justification: Economies of Worth*. Princeton, NJ: Princeton University Press.

Brown, P., Lauder, H., and Ashton, D. (2011). *The Global Auction: The Broken Promises of Education, Jobs and Incomes*. Oxford: Oxford University Press.

Brown, W. (2015). *Undoing the Demos: Neoliberalism's Stealth Revolution*. Brooklyn, NJ: Zone Books.

Campaign for Social Science (2015). *The Business of People: The Significance of Social Science over the Next Decade*. London: Academy of Social Sciences and Sage. Retrieved 5 January 2017 from: http://campaignforsocialscience.org.uk/wp-content/uploads/2015/02/Business-of-People-Full-Report.pdf.

---

[7] This chapter was written as the Higher Education and Research Bill 2016–2017 was passing through Parliament and had reached its final committee stage in the House of Lords. The Bill would provide separate regulatory arrangements for teaching and research, allow the use of university title by single-subject, teaching-only, for-profit providers, and facilitate the entry of new providers and 'exit' from the 'market' and place. Regulatory arrangements would be under the direct administration of the Minister of State without Privy Council oversight. For a critique, see Holmwood et al. (2016).

Casey, L. (2016). *The Casey Review: A Review into Opportunity and Integration*. London: Department for Communities and Local Government. Retrieved 5 January 2017 from: www.gov.uk/government/publications/the-casey-review-a-review-into-opportunity-and-integration.

Collini, S. (2012). *What are Universities For?* London: Penguin.

Committee on Higher Education (1963). *Higher Education: Report of the Committee Appointed by the Prime Minister under the Chairmanship of Lord Robbins 1961–63* [the 'Robbins Report'], Cmnd 2154. London: Her Majesty's Stationery Office.

Dewey, J. (1927). *The Public and its Problems*. Athens, OH: Ohio University Press.

Etzkowitz, H. (2008). *The Triple Helix: University-Industry-Government Innovation in Action*. New York: Routledge.

Friedman, M. (1962). *Capitalism and Freedom*. Chicago, IL: University of Chicago Press.

Gibbons, M., Limosges, C., Nowotny, H., Schwartzmann, S., Scott, P., and Trow, M. (1994). *The New Production of Knowledge*. London: Sage.

Ginsberg, B. (2011). *The Fall of the Faculty: The Rise of the All-Administrative University and Why It Matters*. New York: Oxford University Press.

He, L. (2016). Transnational higher education institutions in China: A comparison of policy orientation and reality. *Journal of Studies in International Education*, 20(1), 79–95.

Holmwood, J. (2011). The idea of a public university. In J. Holmwood (ed.), *A Manifesto for the Public University* (pp. 12–26). London: Bloomsbury Academic.

Holmwood, J. (2016a). 'The turn of the screw': Marketization and higher education in England. *Prometheus: Critical Studies in Innovation*, 34(1), 63–72.

Holmwood, J. (2016b). The university, democracy and the public sphere. *British Journal of the Sociology of Education* [online], 1 September, 1–13, doi: 10.1080/01425692.2016.1220286.

Holmwood, J., and Bhambra, G. K. (2012). The attack on education as a social right. *South Atlantic Quarterly*, 111(2), 392–401.

Holmwood, J., Hickey, T., Cohen, R., and Wallis, S. (2016). *The Alternative White Paper for Higher Education: In Defence of Public Higher Education – Knowledge for a Successful Society*. London: Convention for Higher Education. Retrieved 5 January 2017 from: https://heconvention2.files.wordpress.com/2016/06/awp1.pdf.

Jones, R. (2013). The UK's innovation deficit and how to fix it. Sheffield Political Economy Research Institute, paper no. 6. Retrieved 5 January 2017 from: http://speri.dept.shef.ac.uk/wp-content/uploads/2013/10/

SPERI-Paper-No.6-The-UKs-Innovation-Deficit-and-How-to-Repair-it-PDF-1131KB.pdf.

Kalleberg, A. (2011). *Good Jobs, Bad Jobs: The Rise of Polarized and Precarious Employment Systems in the United States, 1970s to 2000s*. New York: Russell Sage Foundation.

Kerr, C. (2001 [1963]). *The Uses of the University*. Harvard, MA: Harvard University Press (originally published 1963).

Kuznets, S. (1953). *Shares of Upper Income Groups in Income and Savings*. Cambridge, MA: National Bureau of Economic Research.

Lippmann, W. (1925). *The Phantom Public*. New York: Macmillan.

May, T. (2016). Britain after Brexit: A vision of a global Britain. Conservative conference speech. *Politics Home*, posted 2 October. Retrieved 5 January 2017 from: www.politicshome.com/news/uk/political-parties/conservative-party/news/79517/read-full-theresa-mays-conservative.

Mazzucato, M. (2011). *The Entrepreneurial State: Debunking Public vs Private Sector Myths*. London: Anthem Press.

Morgan, J. (2016). The problems of populism: Tactics for Western universities. *Times Higher Education*, 3 November. Retrieved 5 January 2017 from: www.timeshighereducation.com/features/the-problems-of-populism-tactics-for-western-universities.

Müller, J.-W. (2016). *What is Populism?* Philadelphia, PA: University of Pennsylvania Press.

Nussbaum, M. C. (2010). *Not for Profit: Why Democracy Needs Humanities*. Princeton, NJ: Princeton University Press.

OECD (2015). *All On Board: Making Inclusive Growth Happen*. Paris: OECD Publishing. Retrieved 5 January 2017 from: www.oecd-ilibrary.org/development/all-on-board_9789264218512-en.

Oxford Dictionaries (2016). Word of the year. *Oxford Dictionaries*. Retrieved 5 January 2017 from: https://en.oxforddictionaries.com/word-of-the-year/word-of-the-year-2016.

Phillips, T. (2016). China universities must become Communist party 'strongholds', says Xi Jinping. *Guardian*, 9 December. Retrieved 5 January 2017 from: www.theguardian.com/world/2016/dec/09/china-universities-must-become-communist-party-strongholds-says-xi-jinping.

Piketty. T. (2014). *Capital in the Twenty-First Century*. Cambridge, MA: The Belknap Press.

Smith, S. (2011). Afterword: A positive future for higher education in England. In J. Holmwood (ed.), *A Manifesto for the Public University* (pp. 127–142). London: Bloomsbury Academic.

Snowdon, P. (2012). *Sock Puppets: How the Government Lobbies Itself and Why*. London: Institute for Economic Affairs. Retrieved 5 January 2017 from: https://iea.org.uk/wp-content/uploads/2016/07/DP_Sock%20Puppets_redesigned.pdf.

Waterson, P. (2014). *Homes for Local Radical Action: The Position and Role of Local Umbrella Groups*. Working paper no. 7. London: National Coalition for Independent Action Inquiry into the Future of Voluntary Action. Retrieved 5 January 2017 from: www.independentaction.net/wp-content/uploads/2014/08/Role-of-local-umbrella-groups-final.pdf.

Epilogue: publics, hybrids, transparency, monsters and the changing landscape around science

Stephen Turner

Science, science journalism and the academic study of science itself are grappling with rapid changes in the nature of their object in the three disciplines of science and technology studies, philosophy of science, and history of science. Science is no longer the familiar world of laboratories and chalkboards full of equations, with the public at a discreet remove, buffered by a set of benign images of unworldly scientists pursuing arcane interests. These images of science were captured by the Ronald Reagan movie *Bedtime for Bonzo* (de Cordova, 1951), which featured a charming scientist living with a chimpanzee to test his theories about nature and nurture. The science was individualistic, rather than a team effort; the scientist was a sincere and harmless seeker of truth and friend of humanity removed from ordinary concerns and bumbling in his human interactions, and the results were understandable in human terms, though largely without practical significance.

The disciplines studying science, journalists and scientists themselves upheld aspects of this image well into the 1960s. Science was a separate world of special practices and rituals, with an individualistic epistemology, in which science was ultimately a matter, as Michael Polanyi put it, of personal knowledge (1958). In this idealised image of science, it was governed by merit. Funding, self-promotion, and academic and scientific politics and conflict were neatly excised from the image of the scientist. Science occasionally challenged our sense of ourselves as humans, or our place in the universe, but for the most part it was hidden from the public and appeared in popular science magazines and boys' fiction as technology.

Epilogue

Elements of this image have been gradually chipped away, however incompletely, in the disciplines studying science. But scientists themselves, as well as their publics, have now progressed beyond chipping away at the image into consciously transforming it: 'opening' science. The sciences are being opened in many ways. The chapters of this book begin to address this change. As in all large-scale social change, it is difficult for the participants to understand: the old landmarks remain, chalkboards can still be found and laboratories are more impressive than ever, but the landscape itself has become strange.

Like the British landscape itself, the British science landscape is imbued with meaning and ancestral meanings that lurk below the surface, together with a few monsters. The issue of social class was central to the Great Devonian controversy of the early nineteenth century (Rudwick, 1985), which was itself the result of a novel form of public science – survey geology sponsored by the state. By the end of that century Karl Pearson was promoting a vision of modernity shorn of religion, with scientists as the new priests and the state, run by experts, as the new object of veneration (1888: 20, 130–131, 133–134). With it came a model for science education in which workers were taught elementary science with the goal of impressing on them the greatness of science and its authority. By the 1920s, fear of the effects of science led to calls for a moratorium on new discoveries by the Bishop of Ripon (Burroughs, 1927: 32). In the 1930s, under the influence of Marxism, a vast body of popular science writing promoted the idea that the full use of science was central to economic well-being, and that it was being held back by backward capitalist elites working through relations of production that had been made obsolete by the forces of production of modern technology. This was called – the title of one of the key texts of the era – *The Frustration of Science* (Hall, 1935), a concept endorsed by such major figures as Frederick Soddy and John Desmond Bernal.

Where was the public in all this? It was present, surprisingly. The geniuses of science were known to the public, discoveries were celebrated, and science education was initiated in Britain in the form of working men's institutes with lectures on elementary science, and there were best-selling books, such as *Mathematics for the Million* (Hogben, 1937) and the *ABC of Relativity* (Russell, 1925). There was a lively interest in amateur science: radio clubs, for example, sent small donations to support Jodrell Bank, the array of radiotelescopes that was the first major achievement of post-war British science. And this produced a particular inflection on the relation of science and the public – scientists wanted support and deference, and also, especially in the leftist science writing of the 1930s, claimed to have a fundamental solidarity with the working class by ascribing scientific knowledge to craftsmen and identifying scientists with the 'workers'.

These writings, directed at the public, promoted a model of public engagement of science that was not far removed from Pearson's: scientists were workers, but workers whose capacity for rationalising production and economic life generally made them *primus inter pares* in the future planned society, in which science itself would be planned for the public benefit. The plan would be produced collectively by the trade unions, representing the productive units of the plan, coordinated at the top by science. Science itself was understood as technology, a form of rational engineering that could be effectively applied to the problems that had formerly been addressed by markets and politics.

The idea of planning was a fetish of the 1930s. It was opposed by liberal scientists and economists, with the scientists calling for the preservation of the autonomy of science and its freedom. They also rejected the equation of science and technology. In both cases, part of the justification was efficiency – ironically, an idea central to Pearson – that made free markets and free science superior to planned science. And they did not support the idea that science could replace politics. But, in one of those dialectical moments that history occasionally provides, this discussion was upended by a scientific, technological and planned achievement that produced a radically new problem on the status of science and its relation to the public: the atomic bomb project.

The shock of the bomb led to a new sense of the responsibilities of scientists for the consequences of science. The reaction was powerful: scientists organised to assert themselves and sought ways to control the nuclear menace. The anti-nuclear movement in science was paralleled by a powerful social movement, the Campaign for Nuclear Disarmament. This added a new element that simultaneously empowered scientists, placed them in the public eye and demanded something new out of them. But their reaction traded on the same image of science as the font of authority, and the idea that this authority extended to politics, especially to a domain of politics outside the planning paradigm: international politics. The idea was still that scientists, not politicians, should be in control.

This sense evolved in the face of political reality, and new hybrid organisations were created, especially the Pugwash conferences devoted to the eventual elimination of nuclear arms. This still asserted the authority of science. On the website of the organisation we read that 'The mission of the Pugwash Conferences on Science and World Affairs is to bring scientific insight and reason to bear' (Pugwash Conferences on Science and World Affairs, n.d). But it did so by 'meetings and projects that bring together scientists, experts, and policy makers'. This may seem like a minor change, but it was nevertheless important. The older social relations of the science movement, with its reliance on the model of craft unions to organise scientific workers, owed its inspiration to the guild socialism of the 1930s and to the idea of exclusion. Pugwash, which had a highly active UK branch, was based on a recognition that science was not the only voice that needed to be heard in order for scientists to carry out their own responsibilities.

Nuclear weaponry was a novel issue with implications for the future of humanity: a monster. It was an intellectual challenge and a comprehensive one: it challenged basic ideas of security and the meaning of life, as well as the whole notion of warfare and national security. It took a long time for central political institutions as well as their intellectual coteries in the fields of international relations and security studies to even grasp the issues. And the positions that ultimately emerged were not especially friendly to scientists. The topic was so large and complex that it generated counter-expertise, and a whole panoply of actors – scientists speaking as citizens and sometimes politicians, self-proclaimed spokespersons for humanity, social movements, a bomb literature like the war literature of the past, and many more. We now have more monsters and more people who have a stake in science, together with opinions that diverge from those of scientists and more organised movements and currents of thought.

From this point on one can discern two movements: on the one hand, science, in the form of such things as evidence-based decision making, risk analysis and regulatory science, appears in more and more domains. On the other, the practitioners in these new domains are compelled to deal with the people who already occupy them or who have a stake in them and special knowledge about them: the public, policymakers, and practitioners and participants in related social movements or causes. The new landscape was produced by these two directions of change. Science now consisted of much more than laboratories and chalkboards, but there was no guide to this new landscape; its meanings had to be constructed on the fly.

The merit of detailed case studies is that they tell us something new. They tell us where old templates and old expectations are wrong, and complicate the attempt to reduce novel phenomena to a simple formula. In the case studies in this volume we find many complications.

In the first part of this book, on transparency and openness, we can begin to see the larger outlines of the problem. Openness and transparency are ideas that conceal many different and potentially conflicting agendas, even agendas that are entirely opposed, such as the use of the concept to enable the powerful to engage in surveillance as well as to empower the disempowered by giving them access to the knowledge and practices of the powerful in order to monitor them. From the point of view of democracy, this means that transparency can cut in many different ways to diminish the privacy and therefore autonomy of subjects as well as those of the powerful, who use secrecy as a weapon (see Worthy and Hazell, 2013). Open access is an example of the complexities of the consequences of actual policies, as Stephen Curry shows. The simple motivation is to get the taxpayers access to the science they have paid for, and perhaps more quickly than can be done by the traditional journal model of publication. But one startling effect, as he points out, is that the distinctions between real science and something less real are being radically undermined by the rise of rapid response blogging, which blurs the line between academic criticism controlled by the journal system and sheer opinion. He gives the example of a blog response to the questionable claim by NASA to have found microbes on Mars, to which NASA declined to respond. But recent events have shown this blurring to be more pervasive. There is the long-running (and much despised by scientists) practice of blogs criticising the claims of climate science, together with the example of 'methodological terrorism' practised by critics of the work of Princeton psychologist Susan Fiske in blogs, and not through private channels or the journal system (*Association for Psychological Science*, 2016; Letzter, 2016; Singal, 2016). Non-response to criticism is no longer an option, if one's public reputation matters.

Epilogue

Yet there can be strong motivations for transparency even where there are fundamental conflicts. Carmen McLeod shows, in her discussion of the issue of research involving laboratory animals, how openness initiatives in various countries attempted, with partial success, to overcome the distrust of animal-rights activists. As she notes, this is one of the longest running and most intense of all conflicts between science and society, marked by anti-vivisectionist legislation and occasional violence. Here, an organised effort by scientists to open up research has been made, with the aim of reducing opposition and increasing public trust. But cherry-picked information and the kinds of rationalistic utilitarian arguments presented with this information may, as she shows, have no effect on trust itself. Evidence-based policy is one of the mechanisms by which so-called expert practices suspected of arbitrariness can be opened up to scrutiny in contested spheres. But what is evidence? As Roda Madziva and Vivien Lowndes show, in the case of an immigration officer's determinations of persecution, the evidence is itself subject to systemic biases and errors, leading to new kinds of contestation.

In each of these cases, transparency is a double-edged sword. It is assumed that transparency places everyone at the same level. But, in part because of the different viewpoints of those whose activities are being made transparent and of those who are assessing the activities, the result is often the opposite: more contention and less trust. These are, however, cases of hands-off opening up and transparency, without actual engagement, geared more to the relations of science with an abstract public, such as the simple making of scientific publications available through open access. A new dimension is introduced with the attempt to deal with the same kinds of issues through the self-conscious creation of new, hybrid, organisational forms.

Future Earth is a project of exactly this kind: an attempt to bring together different forms of sustainability knowledge and create both a new kind of science and a new social contract between science and society. As Eleanor Hadley Kershaw, in the first chapter of the second part of the book, devoted to responsibility and openness, shows, this project is, however, plagued by ambiguities resulting from its ambition. The extent to which it can be regulated and made coherent and the openness to diversity required by its remit are particularly ambiguous. This points to a problem that becomes apparent in the next part of the book, on expertise: whether it is possible to create such a new science with a new relation to society without the limitations of normal disciplinary knowledge. She suggests that perhaps the tensions and ambiguities are a virtue that will help sustain the project.

Alison Mohr describes a revealing example of another organisational innovation: a project to provide for a transition from traditional energy sources to low-carbon forms of energy in the global South. The SONG project was an attempt to co-design off-grid systems in Bangladesh and Kenya. It was transdisciplinary and offered an active role for social scientists. The framework, nevertheless, was top down. What emerged were unanticipated conflicts over justice and differing aspirations for the future between generations and genders, in a setting in which community inclusion was required. Here, the unexpected conflicts over issues of justice worked to undermine the project itself.

These cases raise a fundamental question about monstrousness, and the monstrousness of the unsolved and unresolved, which animates these conflicts. As Stevienna de Saille and Paul Martin note, the anthropologist Martijn Smits, writing on novel technologies, says that monsters embody 'characteristics that make them both horrifying and fascinating at the same time' (p. 149). This produces simultaneous fear and attraction that can never be reconciled. They are thus especially mythogenic, which becomes apparent in the controversies over genetically modified organisms (GMOs) and the earlier controversy over recombinant DNA and the use of synthetically modified algae as a source of oil to replace palm oil by a natural food company named Ecover. In these cases, engaging the public was not enough: the imaginaries constructed around the products, which were fed by larger social imaginaries, prevented these monsters from being tamed. And the idea of responsible innovation itself may be an attempt to resolve irreducible contradictions intrinsic to the innovations themselves.

One apparent solution to the problem of resistance to innovations is public inclusion in decision-making processes themselves. In the next part, entitled Expertise, Sarah Hartley and Adam Kokotovich deal with an example of this: the formal requirements for public involvement in risk assessment; that is, opening up the risk-assessment process itself. These are not often met in practice. But why? As the authors show, in the case of the European Food Safety Authority, the bureaucratic division of labour purports to separate value issues from science and allocate distinct roles for the public and science. But this means that the implicit value judgements that are part of the science are excluded from public control and scrutiny. And the chapter points to another important phenomenon. Risk assessment itself is a case in which a bridge discipline is created that is not purely scientific and becomes 'disciplinary' in character, with the effect of undermining its openness to the public, despite the formal machinery provided of public involvement.

Judith Tsouvalis's study of ash dieback shows the complexity of the entanglement of science with regulatory and legal regimes, and the way in which these regulatory regimes can be backward looking. This makes it difficult for novel scientific results to have an effect. But the study points to a result similar to that found by Hartley and Kokotovich. Particular bureaucratic and academic specialties with quasi-scientific status, such as 'biosecurity', figure in this case, and these hybrid fields mediate between the more traditional science of plant pathology and an interested public. On the one hand, this novel discipline empowers the public by giving it someone to represent it and its interests. But it does so in ways that disempower the public by narrowing the issues into technical ones and excluding the public's own voices.

The issues we have encountered so far have typically involved conflicts between broader notions of participation and top-down models of public understanding of science, which involve the acceptance of the special authority of science. There is also a tropism toward disciplinarisation. The same tropism is evident in the development of the novel discipline of climate science. But climate science from the start had a policy objective and sought to influence public policy. It became spectacularly successful in this respect. But how? The film *An Inconvenient Truth (AIT)* by Al Gore (Guggenheim, 2006) represents an instructive case. Gore spoke both to, and on behalf of climate science, and for the people to the politicians. But the Gore story presents even more ambiguities. Gore created himself as a new type, a prophet–politician–activist–explainer of science, and the mediatisation of global warming shows new aspects of the relation between the public and science. The public needs, as Gore understood, to be created – to be formed, enlightened and energised as a force. And he succeeded by creating a phrase, 'an inconvenient truth', which, as Warren Pearce and Brigitte Nerlich explain, became instantly recognisable and part of the language. But Gore's efforts also produced a counter-politics and a scientific counterpublic, made possible by the exaggerations that made *AIT* so successful.

Gore's movie is a solution to the problem of including the public. We can think of these cases as failed or only partially successful attempts at hybridisation and openness, but none of them were self-conscious attempts to open science or create a new relation between science and the public. Instead they were attempts to achieve goals or values of different kinds, from different perspectives and sources of knowledge, but in concert with one another, and in doing so they required the invention of particular forms, such as new discipline-like bodies of publicly valuable knowledge, or new templates that could be used to mobilise public sentiment. This differs from a self-conscious pursuit of novel forms of hybridity in the name of openness as a value, or out of a desire to bring science and the public together under a novel, organised intellectual structure.

But what is the public? Sujatha Raman, Pru Hobson-West, Mimi E. Lam and Kate Millar approach this problem through the issue of minority perspectives. As they note, discussions of the pluralisation of the public into unruly and competing perspectives has led to a neglect of the processes by which the public interest is itself constructed. In the case of animal research in the UK, where this is an especially salient issue, the construction involves opinion polls that serve to sideline protesters. The example of Canadian fisheries policy and the Canadian Government's legally adjudicated conflicts with First Nations' interests and conceptions shows how these minority viewpoints are marginalised, but also how they might be inserted into broader discussions of the public good.

What if science itself is the monster? This is the problem posed by the oldest of these conflicts, the conflict between religion and science, explored in the last part of the book, devoted to faith. In the two final chapters the theme is taken up by Fern Elsdon-Baker and by David Kirby and Amy Chambers, who make an interesting point about the ways that this conflict has been both managed and inflated. Elsdon-Baker discusses the construction of creationism as an alien doctrine threatening scientific progress and social progress. But as she shows, few people adhere to the full menu of the creationist doctrine, and ethical concerns over the doctrines of eugenics and practices of forced sterilisation have historically fuelled opposition to a Darwinian world view. The inflated bogey of creationism is a monster that obscures these and other moral concerns. Kirby and Chambers reveal a history in which the Catholic Church in the USA endeavoured to influence the image of science through the mechanisms of quasi-voluntary censorship agreed to by Hollywood. Its concerns were various, but many of them involved the idea that science could replace God or lead to a future without God. It was the opposition of religion to science itself that especially perturbed it: the Church sought instead a kind of accommodation in which the limits of science were acknowledged and a place for religion retained. These chapters show that the binary oppositions between science, which Elsdon-Baker describes as a 'nebulous object', and its various 'others' are unstable and constantly being constructed and reconstructed. This is a lesson of the other chapters as well. Once we look closely at such things as the public, they, too, fragment and disappear.

When Hadley Kershaw alludes in her title to the Leviathan, the Hobbesian monster created by our actions in the form of a mysterious act of collective willing, it is a reminder that science is not the only creator of monsters. The state, with its inhuman power, the 'mortal God', as Hobbes put it, is also is one of those things that has 'characteristics that make them horrible and fascinating at the same time'. Science routinely produces such results: whether it is the horrors of the treatment of laboratory animals or the idea that we have inadvertently, through our normal daily actions, produced the collective catastrophe of anthropogenic global warming, science is a special and prolific source of monsters. Openness and transparency are the supposed cure for this sense of the monstrous: they make science less alien. We can tame these monsters by turning them into ordinary objects of expert knowledge; for example, by making risk into the sort of thing that can be subjected to a bureaucratic regime, with a hybrid discipline of risk assessment and a special provision for public input. Or we can mediate the oppositions that give rise to the monsters and which the monsters in their dual nature embody by constructing and reconstructing them, or by creating social and epistemic movements, such as Future Earth, in which the oppositions become prosaic organisational problems. But in the end these devices create new monsters that are as alien as the old ones.

References

Association for Psychological Science (2016). Draft of *Observer* column sparks strong social media response. *Observer*, 21 September. Retrieved 1 August 2017 from: www.psychologicalscience.org/publications/observer/obsonline/draft-of-observer-column-sparks-strong-social-media-response.html.

Burroughs, E. A. (1927). Is scientific advance impeding human welfare? *Literary Digest*, 95, 32.

de Cordova, F. (1951). *Bedtime for Bonzo*. Universal City, CA: Universal International Pictures.

Guggenheim, D. (2006). *An Inconvenient Truth*. Los Angeles: Paramount Classics.

Hall, D. (1935). *The Frustration of Science*. London: George Allen and Unwin.

Hogben, L. (1937). *Mathematics for the Million*. New York: W. W. Norton.

Letzter, R. (2016). Scientists are furious after a famous psychologist accused her peers of 'methodological terrorism'. *Business Insider*, 22 September. Retrieved 24 October 2016 from: www.businessinsider.com/susan-fiske-methodological-terrorism-2016-9.

Pearson, K. (1888). *The Ethic of Freethought: A Selection of Essays and Lectures*. London: T. Fisher Unwin.

Polanyi, M. (1958). *Personal Knowledge: Towards a Post-Critical Philosophy*. Chicago: University of Chicago Press.

Pugwash Conferences on Science and World Affairs (n.d.). *Pugwash Conferences on Science and World Affairs*. Retrieved 17 October 2016 from: https://pugwash.org/.

Rudwick, M. J. S. (1985). *The Great Devonian Controversy: The Shaping of Scientific Knowledge among Gentlemanly Specialists*. Chicago: University of Chicago Press.

Russell, B. (1925). *ABC of Relativity*. London: George Allen and Unwin.

Singal, J. (2016). Inside psychology's 'methodological terrorism' debate. *New York Magazine*, 12 October. Retrieved 24 October 2016 from: http://nymag.com/scienceofus/2016/10/inside-psychologys-methodological-terrorism-debate.html.

Smits, M. (2006). Taming monsters: The cultural domestication of new technology. *Technology in Society*, 28(4), 489–504.

Worthy, B., and Hazell, R. (2013). The impact of the Freedom of Information Act in the UK. In N. Bowles, J. T. Hamilton and D. A. Levy (eds), *Transparency in Politics and the Media: Accountability and Open Government* (pp. 31–45). London: L. B. Tauris.

Index

accountability 8, 23, 24, 36, 39, 57, 69, 110, 111, 121, 122, 181, 234, 236
animal research 8, 30, 43, 55–58, 60–66, 68–70, 170, 174, 233, 238–242, 244, 330
animal rights 56, 58–60, 65, 239–241, 246, 327
anti-evolution 259, 261–263, 267, 269–271, 273
anti-science 152, 153, 255, 269, 273, 293
Assange, Julian 25
asylum 8, 30, 75–78, 80
authority 3, 43, 47, 88, 117, 177, 187, 235, 254, 282, 315, 323–325, 328, 329
automation 1

bacteria 43, 154, 156
Bentham, Jeremy 25, 28, 55
biosecurity 170, 171, 195, 198–202, 205–208, 329
Blair, Tony 173, 230–233, 235, 236, 238, 240, 241, 245–246
blasphemy 78, 85, 86, 284, 285, 293
bodies 23, 26–28, 59, 80, 100, 107, 133, 149, 176, 178, 180, 198, 234, 257, 266, 273, 283, 304, 310, 311, 323, 330
Bok, Sissela 23, 25, 28
Brexit 1, 11–13, 15, 16, 304

Britain 5, 11–13, 15, 16, 26, 45, 46, 56, 61, 67, 80, 90, 102, 195–198, 200, 201, 230, 235, 236, 239, 241, 242, 245, 253, 260, 265, 278, 289, 305, 306, 323, 324
Bush, George W. 174, 218, 220
business 41, 45, 61, 112, 122, 148, 155, 239, 311

Cameron, David 12, 79
capitalism 98, 315, 317, 323
censorship 10, 257, 279–289, 293, 296–297, 331
centralisation 9, 99, 108, 113, 118, 123, 135
China 137, 305
Christianity 16, 30, 77, 78, 80, 256, 257, 279, 280, 287, 288, 290–297
citizen-science 45–47, 102
climate change 6, 172, 212–215, 217, 219, 222, 223
Climategate 2, 6
climate justice 140
climate science 212, 213, 215, 216, 218, 220, 222, 223
Clinton, Hillary 12, 13
confidentiality 8, 25, 30
Congress (US) 158, 218, 220
consensus 14, 15, 23, 99, 116, 123, 214, 304
consent 213, 270

conservatives 12, 16–17, 173, 221, 253, 256, 292, 293, 311
constituencies 17, 40, 172, 175
counterpublic 213, 216, 220, 221, 238, 241, 330
counter-representation 173, 213
creationism 10, 16, 17, 253–267, 270–273, 292, 330, 331
credibility 15, 24, 82, 85, 88, 182, 190, 221, 273
critique 38, 43, 118, 221, 234, 239, 255
cybercrime 1
cyborg 107

deliberation 116, 121, 171, 183
deliberative 121, 175, 184
democracy 10, 17–18, 23, 25, 98, 169, 188, 202, 233, 306, 313, 314, 316, 317, 326
Democrat 212, 223
democratisation 4, 114, 123, 208, 308
Dewey, John 173, 213–216, 220, 222, 223, 307, 312–318
dialogue 49, 64, 66, 69, 99, 116, 187, 195, 205, 235, 238, 284, 290, 314
digital 2, 6, 35, 42, 98, 101
digitalisation 1
dinosaurs 292, 293
direct action 58
disciplines 16, 36, 45, 49, 100, 102, 107, 108, 112, 121, 183, 189, 260, 322, 323, 329, 331
disclosures 26, 29, 61
discourse 23, 28, 57, 64, 78, 79, 89, 90, 104, 178, 201, 202, 241, 246, 259, 264, 265, 267, 268, 270, 271, 273, 304
dissemination 35, 99, 102, 120, 214, 216, 280, 292, 293, 296
dissent 2, 150, 154, 202, 213, 303
diversity 9, 76, 78, 79, 84, 87, 89–90, 99–101, 113, 115, 123, 125, 136, 237, 255, 256, 327

Ebola 41, 42
economics 1, 232, 234, 314

economists 1, 35, 203, 324
ecosystem 112, 200, 242–244, 310
EFSA (European Food Standards Agency) 157, 177, 185–188, 190
energy 1, 7, 9, 42, 98, 131–145, 150, 151, 159, 160, 328
energy transitions 101, 131–133, 137, 143
enlightenment 170
epidemics 172, 199–201, 207–208
epistemology 202, 273, 322
ethics 6, 28, 29, 62, 68, 296
ethnography 16, 253, 255
eugenics 257, 270–272, 281–283, 296, 330
evangelical 16, 83, 85
evidence 2, 4, 6, 8, 15, 27, 30, 69, 75–77, 79, 81–85, 87–90, 97, 116, 176, 177, 179, 184, 188, 203, 205, 208, 212, 220, 223, 237, 244, 269, 271, 274, 304, 307, 325, 327
evolution 10, 16, 253–257, 259–267, 269–274, 285, 286, 292–294, 296
evolutionists 261, 265
experimentation 27, 55, 61, 67, 133, 144, 240, 290
experiments 25, 27, 28, 30, 44, 55–57, 61, 65–67, 113, 124, 136, 137, 145, 169, 230, 237, 238, 285
experts 8, 10, 75, 82, 87–89, 169–171, 175, 178, 181, 183, 184, 187, 189, 195, 198, 202, 204, 205, 212–213, 215, 218, 221, 222, 260, 313, 317, 327, 331
expertise 1, 3, 5–7, 9, 10, 17–18, 76, 82, 84, 99, 102, 121, 141, 167, 169, 170, 172, 174, 180, 189–191, 198, 212–216, 218, 220, 222–224, 234, 246, 258, 313, 317, 327, 328

faith 3, 8, 10, 75–79, 82–84, 86, 87, 90–91, 251, 253, 254, 258, 260, 265, 280, 287, 290, 294, 295, 306, 330
fake asylum claimants 85, 87
'fake news' 6
Faust 25

Index

fear 4, 14, 38, 40, 55, 58, 66, 68, 81, 82, 85, 86, 90, 104, 108, 149–154, 157, 159, 160, 202, 273, 286, 323, 328
filmmakers 10, 279, 285–290, 292, 294–296
Frankenstein 107, 150, 152, 153, 158, 160, 161, 278, 284
freedom 24, 26, 38, 39, 118, 122, 160, 289, 305, 324
funders 30, 36, 37, 39, 41, 42, 63, 67, 99, 111, 112, 116, 239, 240
funding 11, 13–16, 37, 41, 48, 67, 101, 109, 113, 149, 198, 303–305, 307, 309, 311, 322

Gallup 265, 266
Galton, Francis 270, 271
gender 107, 139, 142, 143, 145, 268, 269, 272, 328
Genesis 259
genetics 2, 9, 43, 46, 150, 154, 156, 157, 179, 186, 187, 208, 222, 230, 270, 272, 312, 328
globalisation 201, 304
global North 131, 136, 137, 144
global South 137, 144, 145
God 151, 153, 253–255, 263, 266, 267, 278, 284, 289, 331
Gove, Michael 305
governance 7, 36, 57, 61, 67, 108–111, 119, 121, 135, 141, 143, 145, 149, 162, 170, 176–178, 180, 182–186, 188, 190–191, 201, 206, 233, 235, 236, 245, 260, 303, 309

Habermas, Jürgen 5
Haida Nation 174, 238, 241–244, 246–247
half-animal 69
Haraway, Donna 3, 4, 107, 230, 238, 241
HEFCE 37, 38
hegemony 173, 214–216, 221
herring 242–244
hierarchy 23, 87, 145, 254, 316
Hitchcock, Alfred 287
HIV 42, 270

Hobbes, Thomas 107, 331
Hollywood 46, 153, 280, 286, 288, 291, 292, 295, 331
humanism 273
humanities 14, 38, 39, 110, 112, 208, 309
Huntington Life Sciences 58

identity 44, 78–82, 88–90, 111, 113, 116, 119, 120, 172, 174, 175, 196, 202, 260, 262, 263, 266, 268, 269, 273
ideology 75, 307, 315, 316
ignorance 30, 41, 158, 159, 181
immigration 8, 15, 75–82, 87–90, 327
inclusion 173, 232, 233, 238, 240, 242, 282, 317, 328
income 132, 139, 144, 145, 308, 309
inequality 98, 102, 104, 122, 136, 138, 139, 141, 143, 145, 307–309, 311, 312
intelligence 6, 16, 206, 264, 272
internet 35, 42, 44–46, 49, 98, 292
interpretation 23, 76, 77, 82, 100, 102–104, 110, 115, 123, 184, 237, 257, 284, 287, 296
Irwin, Alan 7, 103, 110, 124, 176
Islam 79, 86, 89, 90, 265, 272
Islamophobia 80, 90, 272
Israel 293

Jasanoff, Sheila 4, 7, 55, 57, 110, 113, 125, 149, 178, 216, 217, 232, 234, 237, 246
journalism 1, 322
journalists 4, 6, 38, 49, 154, 246, 322
Jurassic Park 291–293
justice 7–9, 100, 101, 131–134, 137–141, 143–145, 258, 305, 306, 328
justification 15, 36, 57 174, 285, 303, 307, 324

Kansas 16–18
Kubrick, Stanley 289

Latour, Bruno 107, 152, 160, 195, 202
laypeople 5
legitimacy 3, 10, 26, 75, 141, 149, 169, 170, 174, 177, 190
leviathans 107, 108, 113, 114, 116, 117, 119, 123, 331
liberalism 315

markets 11, 36, 133, 148, 303, 307, 309, 312, 324
May, Theresa 11, 12, 16, 25, 306
medical 26, 42, 45, 46, 61–63, 67, 239–241, 246, 280–283, 295
migrants 15, 77, 80, 270, 306
migration 3, 78, 79
military 157, 202
Mill, J. S. 169
mistrust 57, 69
moderates 16
moderation 305
modernisation 24
modernism 149
modernity 152, 153, 323
monsters 2–4, 7, 9, 55, 69, 75, 77, 85, 90, 103, 104, 113, 131, 148–155, 157–162, 207, 212, 216, 223, 230, 256, 257, 259, 278, 279, 284, 322, 323, 325, 328, 330
monstrare 3
Monstropolis 150, 160
morality 257, 278, 279, 281–283, 287, 289
Muslim 15, 78–81, 84–87, 89–90, 93, 272
myth 152, 239
mythogenic 328
mythology 25

NASA 30, 43, 44, 326
nativism 13, 17, 306
neo-liberalism 3, 11, 24, 201, 208, 303–305, 307, 309–311, 313
networks 79, 108, 124, 125, 136, 174, 234, 240, 242
neurotechnology 42
non-academic 109, 110, 113, 121, 310

non-belief 263
non-Christians 91
non-scientists 253, 254
norms 24, 39, 41, 42, 103, 124, 222, 241
nostalgia 217

Obama, Barack 174
open access 7, 8, 35–42, 44, 46–48
open science 39, 49, 55, 66, 99–101
open-source 101, 171, 198

Panama 26
paradigm 207, 272, 325
peer-review 37, 40, 122, 223
pluralism 100, 306
plurality 43, 44, 76, 84, 90, 116, 117, 118, 121, 122, 125, 135
Polanyi, Karl 322
polarisation 160, 162, 222, 305, 309
policies 17, 26, 28, 35, 37, 38, 79, 98, 185, 187, 231, 245, 272, 281, 283, 286, 303, 304, 309, 312, 326
policymakers 2, 5, 6, 14, 75, 76, 112, 148, 189, 237, 238, 307, 312, 326
policymaking 2, 3, 5, 7, 75, 176, 195, 199, 200, 202, 204, 208, 230, 239
politicians 6, 24, 25, 27, 29, 45, 173, 212, 218, 221, 325, 329
politicisation 2
pollution 45, 47, 140
populism 1, 10, 11, 169, 175, 303–307, 311, 312
privacy 8, 25, 29, 30, 326
privatisation 5, 6, 307, 311
Protestants 151, 254, 256, 258, 279, 287, 290
protests 58, 59, 156, 281, 293
psychiatry 257, 286, 287, 296
public attitudes 266
public engagement 190, 238, 260
public interest 2, 4, 10, 27–28, 37, 40, 42, 44, 158, 170, 172–175, 230–246, 312, 315, 330
publicity 25, 55, 71, 169, 170, 238
publicness 2, 6

Index

public opinion 239
publics 3–6, 8–11, 16–17, 45, 49, 62, 69, 101, 103, 104, 159, 170, 172, 173, 176–178, 181, 195, 197, 230–234, 236–240, 245–246, 254, 255, 257, 258, 260, 263–265, 270, 274, 304, 307, 314, 316, 317, 322, 323
public space 259, 264, 267

rational 24, 27, 29, 75, 76, 124, 160, 259, 296, 324
rationalising 324
rationalistic 327
RCUK (Research Councils UK) 37, 40
reductionism 117, 259, 261, 271, 274
REF (research excellence framework) 37–39, 198
referendum 11–13, 17, 304, 311
regulations 4, 7, 25, 26, 57, 58, 148, 161, 185, 201, 206
regulators 170, 171, 315
relativism 268
relativity 323
representation 172–174, 212–216, 218, 220–223, 238, 259
representatives 36, 156, 172, 303
reproducibility 15
Republicans (US) 11–13, 15–17, 306
rhetoric 75, 76, 79, 169, 272, 312
rights 30, 35, 36, 39, 42, 45, 56, 59, 60, 78, 121, 174, 241–243, 246, 308, 309
risk 7, 9, 10, 13, 15, 81, 83, 100, 104, 140, 154, 155, 161, 170, 171, 176–191, 197, 201–204, 206–208, 265, 307, 325, 328, 329, 331
rituals 291, 322
RRI (Responsible Research and Innovation) 7, 36, 43, 48, 103, 104, 109, 149, 159
rules 25, 40, 61, 122, 123, 136, 187, 197, 283, 311

science–society 1, 55, 56, 58, 64, 66, 109, 110, 113, 230, 232, 233, 236, 245

scientisation 4
Sci-Hub 42
scrutiny 2, 4, 23, 90, 97, 170, 174, 178, 181, 186, 188, 195, 199, 202, 214, 216, 233, 236, 245, 307, 327, 329
secrecy 8, 25, 26, 28, 55–57, 326
secularisation 268, 272
secularism 273
security 2, 8, 15, 25, 28–30, 56, 150, 170, 201, 202, 325
sharing 2, 36, 65, 66, 100, 171, 221, 254
Snowden, Edward 26, 28, 29
socialisation 25
social movement 236
social science 57, 121, 144, 263, 268, 280
solar energy 132, 139–140, 142
syphilis 281, 285

taboos 25
tax 17, 26, 36
taxonomy 196
taxpayers 35, 45, 304, 326
teaching 6, 13, 16, 86, 198, 265, 304
technocracy 10, 134, 171, 175, 204
'technocratisation' 5
technologies 1–3, 6, 7, 9, 57, 107, 137, 142, 148, 152, 153, 161, 176, 177, 202, 208, 219, 239, 281, 282, 328
terrorism 1, 62, 79, 80, 240, 326
Thévenot, Laurent 307
transdisciplinarity 108–110, 120, 124
transparency 2–4, 6–8, 15, 21, 23–29, 36, 43, 44, 56–58, 60–62, 64–69, 75–77, 79, 88, 90, 97, 98, 102, 169, 186, 190, 235, 236, 258, 322, 326, 327, 331
Trump, Donald 6, 11–17, 230, 304
trust 2, 5, 23, 24, 40, 42–44, 55–58, 61, 62, 66–69, 75–76, 80, 90, 148, 158, 160, 161, 174, 195, 199, 223, 235, 241, 270, 273, 317, 327
truth 6, 10, 49, 85, 172, 212–214, 219, 221, 263, 273, 304, 322, 329
truthfulness 82

UNHCR (United Nations High Commissioner for Refugees) 78, 81, 82, 88, 93
universities 5, 6, 13, 35, 36, 59, 62, 63, 67, 198, 304–307, 311

values 69, 170, 171, 178–180, 183, 184, 188, 190, 268
violence 56, 58–60, 139, 153, 240, 288, 291, 327
vivisection 56, 67

welfare 56, 64, 311
whistleblowing 26, 28

Whitehead, A. N. 61
WikiLeaks 26, 28, 29, 102
Willetts, David 35
women 13, 77, 98, 140, 142, 151, 280, 283
Wynne, Brian 5, 55, 109, 113, 125, 149, 154, 166, 176, 180, 186, 232, 233, 243

Zika 2, 41, 42
zombies 153
Zooniverse 47